FE MECHANICAL STUDY GUIDE

Fe Mechanical Exam Prep: The Most Updated Study Guide with 1500+ Realistic Practice Problems, Diagrams, Step-by-Step Solutions, and a Study Plan. Includes Tips and Tricks for Faster Problem Solving.

Jeremy Crainstone

© 2024 FE Mechanical Study Guide All rights reserved. This book is for informational use only. The publisher is not liable for any damages from its use. Unauthorized copying or sharing, in full or part, is forbidden. Trademarks and brand names are owned by their respective holders. This book is provided 'as is' without any warranties, express or implied.

Table of Contents

SCAN THE QR CODE: 9

Overview of the Exam 10
    What is the FE Exam? 10
    Importance of the FE Credential 10
    Exam Format and Question Types 11
    Efficient Use of FE Reference Handbook 12

Chapter 1: Mathematics 14
    Analytic Geometry Basics 14
    Calculus: Differential and Integral Concepts 15
    Ordinary Differential Equations 16
    Linear Algebra in Engineering 18
    Numerical Methods and Error Analysis 19
    Algorithm and Logic Development 20

Chapter 2: Probability and Statistics 22
    Probability Distributions and Applications 22
    Central Tendency and Dispersion Measures 23
    Expected Value 24
    Regression and Curve Fitting 25

Chapter 3: Ethics and Professional Practice 27
    Codes of Ethics in Engineering 27
    Public Health, Safety, and Welfare 28
    Intellectual Property in Engineering 29
    Societal Impact of Engineering 31

Chapter 4: Engineering Economics 33
    Time Value of Money Concepts 33

    Cost Types and Breakdowns 34

    Economic Analyses 35

Chapter 5: Electricity and Magnetism 37

    Electrical Fundamentals 37

    DC Circuit Analysis 37

    AC Circuit Analysis 38

    Motors and Generators Basics 40

Chapter 6: Statics 42

    Resultants of Force Systems 42

    Concurrent Force Systems 43

    Equilibrium of Rigid Bodies 44

    Frames and Trusses Analysis 45

    Centroids and Moments of Inertia 46

    Static Friction Concepts 48

Chapter 7: Dynamics, Kinematics, Vibrations 50

    Kinematics of Particles 50

    Kinetic Friction 51

    Newton's Second Law for Particles 52

    Work-Energy of Particles 53

    Impulse-Momentum of Particles 54

    Kinematics of Rigid Bodies 55

    Kinematics of Mechanisms 57

    Newton's Second Law for Rigid Bodies 58

    Work-Energy Methods for Rigid Bodies 59

    Impulse-Momentum in Rigid Body Dynamics 60

    Free and Forced Vibrations Analysis 62

Chapter 8: Mechanics of Materials 64

    Shear and Moment Diagrams 64

Stress Transformations and Mohr's Circle 65

Stress and Strain from Axial Loads 66

Stress and Strain from Bending Loads 68

Torsional Stress and Strain Effects 69

Stress and Strain from Shear Forces 70

Thermal Stress and Strain 71

Combined Loading Analysis 73

Deformation Calculation Methods 74

Column Buckling Behavior 75

Statically Indeterminate Systems 77

Chapter 9: Material Properties & Processing 79

Material Properties Overview 79

Stress-Strain Diagrams 80

Ferrous Metals: Properties and Uses 81

Nonferrous Metals: Properties & Applications 82

Engineered Materials Overview 84

Manufacturing Processes 85

Phase Diagrams and Heat Treatment 86

Materials Selection Criteria 87

Corrosion Mechanisms and Control 88

Failure Mechanisms and Prevention 90

Chapter 10: Fluid Mechanics 92

Fluid Properties 92

Fluid Statics and Hydrostatic Forces 93

Energy, Impulse, and Momentum Principles 94

Internal Flow in Pipes and Ducts 95

External Flow: Boundary Layers and Drag 96

Compressible Flow 97

Power and Efficiency in Fluid Systems 98

  Performance Curves for Pumps and Fans 99

  Scaling Laws for Fans and Compressors 100

Chapter 11: Thermodynamics 102

  Properties of Ideal Gases and Substances 102

  Energy Transfers in Thermodynamics 103

  Laws of Thermodynamics 104

  Thermodynamic Processes and Energy Changes 105

  Performance of Thermodynamic Components 106

  Power Cycles: Rankine, Brayton, Otto 107

  Refrigeration and Heat Pump Cycles 109

  Nonreacting Gas Mixtures 110

  Psychrometrics in HVAC Applications 111

  HVAC Processes and Systems 113

  Combustion Reactions and Products 114

Chapter 12: Heat Transfer 116

  Conduction: Heat Transfer in Solids 116

  Convection Principles and Heat Transfer 117

  Radiation: Heat Transfer by Waves 118

  Transient Heat Transfer Analysis 119

  Heat Exchangers Design and Analysis 120

Chapter 13: Measurements and Controls 122

  Sensors and Transducers 122

  Control Systems Basics 123

  Dynamic System Response 124

  Measurement Uncertainty 125

Chapter 14: Mechanical Design Analysis 127

  Stress Analysis of Machine Elements 127

  Failure Theories and Analysis 128

- **Deformation and Stiffness** **130**
- **Springs: Design and Analysis** **131**
- **Pressure Vessels and Piping Safety** **132**
- **Bearings: Types and Design Principles** **133**
- **Power Screws: Load and Torque Requirements** **134**
- **Power Transmission Elements** **135**
- **Joining Methods** **137**
- **Design for Manufacturability** **138**
- **Quality and Reliability Principles** **139**
- **Key Components in Mechanical Systems** **140**
- **Engineering Drawing and GD&T Basics** **141**

READ THIS PAGE CAREFULLY BEFORE YOU START PREPARING FOR THE EXAM

By purchasing this book, you gain access to a wealth of additional resources designed to help you prepare effectively. Don't think of your purchase as limited to the physical book in your hands—take advantage of all the extra resources available to you, such as:

- A study plan and topic checklist (crucial to ensure you cover everything required for the exam).
- In-depth theoretical explanations of key exam topics.
- Videos to help you master problem-solving techniques.
- Online and PDF practice test simulations.
- Tutorials on how to use the calculator efficiently and much more!

After consulting with many students who failed their exams, we identified a common issue: THESE STUDENTS WERE NOT ACCUSTOMED TO TAKING THE EXAM ONLINE OR USING THE REFERENCE HANDBOOK. This is a critical problem, as the exam is computer-based, meaning you must train under similar conditions to the actual exam. Otherwise, your chances of passing are significantly reduced.

FOR THIS REASON, WE CHOSE NOT TO INCLUDE PRACTICE EXERCISES IN THE BOOK but instead provide them in digital format. This approach will help you familiarize yourself with the digital exam format. While this may feel challenging at first, it greatly increases your chances of success.

Through conversations with numerous students, we learned the following:

- Most students who failed the exam reported focusing mainly on theory while neglecting practical exercises.
- Most students who passed the exam on their first attempt reported completing a large number of practical exercises and spending less time on theory.

WE RECOMMEND FOLLOWING THE BONUS RESOURCES IN THIS ORDER:

PHASE ONE

1. Scan the QR code and explore all the additional resources we have provided alongside this book.
2. Download the study plan to your computer.
3. Start watching the videos we've made available. Even if you feel lost at first, keep watching and focus on understanding the problem-solving methods demonstrated in each video.
4. Begin using the Reference Handbook immediately. It's essential to familiarize yourself with its structure and content.
5. Watch the calculator tutorials and memorize all the recommended strategies.

PHASE TWO

1. Choose which manual to begin with for each topic. (You'll find additional manuals among the supplementary resources, all of which are excellent for exam preparation.) You're not required to start with this book—you can use the others as well. This book is intentionally more concise and focused because your priority should be solving practical problems.

The key is to leverage all the supplementary materials to ensure comprehensive exam preparation. If you're short on study time or want to pass the exam quickly, we recommend focusing on the videos and additional resources, as they are specifically tailored to practical problems.

PHASE THREE

Start practicing problems yourself!

Happy studying!

SCAN THE QR CODE:

Overview of the Exam

What is the FE Exam?

The FE Exam, formally known as the Fundamentals of Engineering Examination, is a critical step for aspiring engineers on the path to professional licensure. Administered by the National Council of Examiners for Engineering and Surveying (NCEES), this exam assesses the candidate's comprehension and capability to apply key concepts and technical knowledge fundamental to the practice of engineering. It is designed to ensure that those entering the profession possess a minimum competency in their respective engineering discipline, in this case, mechanical engineering. The FE Mechanical exam specifically evaluates understanding in areas such as mathematics, mechanics, material properties, and energy systems, among others, which are pivotal for a successful career in mechanical engineering.

Passing the FE Exam is the first step in obtaining an Engineer-in-Training (E.I.T.) or Engineering Intern (E.I.) certification, which are precursors to pursuing a Professional Engineer (P.E.) license. The significance of the FE credential cannot be overstated; it is often a prerequisite for engineering jobs, enhances the individual's marketability in a competitive job market, and is a testament to the engineer's dedication to their professional development and adherence to industry standards.

The exam format has evolved over time, with the latest iteration being a computer-based test that includes a range of question types such as multiple choice, point and click, drag and drop, and fill in the blank, designed to test the examinee's critical thinking and problem-solving skills in a time-efficient manner. The FE Mechanical exam is meticulously structured to cover the breadth of knowledge essential in the mechanical engineering field, ensuring that candidates are well-rounded and prepared for the challenges of the profession.

Understanding the FE Exam's structure, content, and objectives is crucial for candidates aiming to pass the exam. With focused preparation and a thorough grasp of the exam's core topics, aspiring engineers can set a solid foundation for their careers, opening doors to advanced opportunities and professional recognition in the field of mechanical engineering.

Importance of the FE Credential

Achieving the FE (Fundamentals of Engineering) credential signifies a pivotal milestone in an engineer's career, underscoring not only a robust understanding of fundamental engineering principles but also a commitment to the profession's ethical and technical standards. This certification, administered by the National Council of Examiners for Engineering and Surveying (NCEES), is recognized across the United States as a prerequisite for pursuing a Professional Engineer (PE) license. The FE credential is highly regarded by employers and peers alike, serving as a testament to the holder's dedication to their personal and professional development.

For engineers, the FE certification opens doors to career advancement opportunities that are often not accessible to those without it. In many engineering firms and government positions, having an FE certification is a requirement for promotion to more responsible roles, including project management and senior engineering positions. This is because the certification is seen as evidence of the individual's comprehensive knowledge of engineering fundamentals, their ability to apply this knowledge to solve complex problems, and their adherence to recognized professional standards.

Moreover, the FE credential enhances an engineer's professional credibility. In the eyes of clients, colleagues, and the broader community, it establishes the engineer as a competent and reliable professional who is committed to upholding the highest standards of engineering practice. This credibility is crucial for building trust and can significantly impact an engineer's ability to win contracts, lead projects, and influence decisions within their field.

In addition to career and credibility benefits, preparing for and passing the FE exam itself offers immediate value. The process encourages engineers to review and solidify their understanding of key concepts and practices within the mechanical engineering discipline, ensuring they remain sharp and current in their knowledge. This rigorous preparation not only aids in passing the exam but also in enhancing the engineer's overall competency and confidence in their professional abilities.

Therefore, obtaining the FE certification is more than just an academic achievement; it is a strategic career move that positions engineers for success in their pursuit of licensure, enhances their marketability, and solidifies their reputation as committed and competent professionals in the field of engineering.

Exam Format and Question Types

The FE Mechanical exam is a comprehensive assessment designed to evaluate a candidate's knowledge and understanding of key concepts essential for a career in mechanical engineering. The exam is structured into a computer-based format, comprising approximately 110 questions that must be completed within a 6-hour session. This time frame includes a scheduled break and time for a brief tutorial on the exam format at the beginning of the session. The questions are predominantly multiple-choice, with each question having four possible answers from which the candidate must select the most appropriate one.

In addition to multiple-choice questions, candidates may encounter alternative question types such as point and click, drag and drop, and fill in the blank. These question formats are integrated into the exam to assess a candidate's ability to apply theoretical knowledge to practical, real-world scenarios. For instance, a point and click question may require the examinee to identify a specific area on a diagram that corresponds to the correct answer, while a drag and drop question might involve arranging given elements into the correct sequence or configuration.

The distribution of questions across the exam topics is aligned with the NCEES FE Mechanical exam specifications. These specifications outline the percentage of questions dedicated to each subject area, ensuring that the exam comprehensively covers the breadth of knowledge expected of a competent, entry-level mechanical engineer. Subjects include but are not limited to mathematics, mechanics, material properties, thermodynamics, and fluid mechanics, among others.

Candidates are advised to familiarize themselves with the FE Reference Handbook provided by NCEES, as it is the only reference material allowed during the exam. The handbook contains equations, figures, and tables that are essential for solving the exam questions. Efficient use of the handbook can significantly enhance a candidate's ability to locate information quickly and accurately during the exam.

By scanning the QR code present in the book, you can access a convenient study plan that is tailored to guide your preparation.

Efficient Use of FE Reference Handbook

Mastering the FE Reference Handbook is crucial for success on the FE Mechanical exam. This comprehensive guide is the only resource allowed during the exam, making it essential to familiarize yourself with its layout, content, and how to navigate it quickly under exam conditions. The ability to efficiently locate equations, tables, and figures can significantly reduce the time spent on each question, allowing for a more thorough review of your answers and a less stressful testing experience.

First and foremost, become adept at using the search function, typically activated by the CTRL+SHIFT+F keyboard shortcut. This function allows you to quickly find specific terms or equations within the handbook. For instance, if you need to find information on the modulus of elasticity, typing this term into the search bar will direct you to the relevant section without the need to manually flip through pages. It's important to practice this skill in advance, as familiarity with the handbook's terminology will expedite your search during the exam.

Understanding the handbook's organization is another key strategy. The FE Reference Handbook is divided into sections corresponding to the major disciplines covered on the exam, with each section containing subtopics arranged in a logical order. Spend time reviewing the table of contents and getting a feel for where each topic is located. Knowing, for example, that the section on thermodynamics precedes the section on fluid mechanics can save precious time during the exam.

Highlighting and tabbing the electronic version of the handbook is not possible during the exam, but you can practice this method during your study sessions. By doing so, you'll develop a mental map of the handbook's layout, which will aid in quick navigation on exam day. Remember, the version of the handbook available during the exam may differ slightly from the one you used to study, so it's wise to regularly check the NCEES website for the most current version and familiarize yourself with any updates or changes.

Another effective strategy is to practice solving problems using only the FE Reference Handbook. This approach not only reinforces your understanding of the handbook's content but also improves your ability to apply this knowledge under exam conditions. During these practice sessions, time yourself to ensure that your search and application methods are both quick and accurate.

Lastly, while the FE Reference Handbook is an invaluable resource, it's important to remember that a deep understanding of the underlying principles and the ability to apply them to solve complex problems is paramount. The handbook should complement your knowledge, serving as a reference for formulas and constants, rather than a crutch. Therefore, a balanced preparation strategy that includes thorough review of exam topics, practice problems, and efficient use of the FE Reference Handbook is essential for success on the FE Mechanical exam.

Chapter 1: Mathematics

Analytic Geometry Basics

Analytic Geometry is a fundamental branch of mathematics that deals with the study of geometric objects through algebraic representation and methods. It primarily focuses on curves, surfaces, and their properties such as perimeter, area, and volume. One of the most significant aspects of analytic geometry is its ability to describe geometric figures using equations, thereby enabling the analysis and solution of geometric problems through algebraic calculations.

Equations of Curves: The equation of a curve is a mathematical relationship that expresses how coordinates x and y are related on a two-dimensional plane. For example, the equation of a circle with radius r and center at the origin $(0,0)$ is given by $x^2 + y^2 = r^2$. Similarly, the equation of an ellipse, parabola, and hyperbola can be represented in standard forms which facilitate the study of their geometric properties and relationships.

Perimeter Calculations: The perimeter of a geometric figure is the total length of its boundary. For standard shapes like triangles, rectangles, and circles, the perimeter can be calculated using well-known formulas. For instance, the perimeter of a rectangle with length l and width w is given by $2(l+w)$, while the circumference of a circle is $2\pi r$, where r is the radius of the circle.

Area Calculations: The area of a shape refers to the size of the surface it covers. The area can be calculated using different formulas depending on the geometry of the shape. For example, the area of a rectangle is lw, the area of a triangle is $\frac{1}{2}bh$ where b is the base and h is the height, and the area of a circle is πr^2. In the context of analytic geometry, the area under curves can be determined using integral calculus, providing a powerful tool for analyzing more complex shapes.

Volume Calculations: Volume measures the space enclosed by a three-dimensional object. For basic solids such as cubes, cylinders, and spheres, simple formulas exist to calculate volume. For example, the volume of a cube with side length a is a^3, the volume of a cylinder with radius r and height h is $\pi r^2 h$, and the volume of a sphere with radius r is $\frac{4}{3}\pi r^3$. Analytic geometry extends these concepts to more complex solids by employing techniques such as triple integration.

Properties of Conic Sections: Conic sections are curves obtained by intersecting a plane with a double right circular cone. They include circles, ellipses, parabolas, and hyperbolas, each defined by specific equations. For example, the general equation for a conic section is $Ax^2 + Bxy + Cy^2 + Dx + Ey + F = 0$, where the values of A, B, C, D, E, and F determine the type of conic. Analyzing these equations reveals important properties such as eccentricity, directrix, and focus, which are crucial for understanding the behavior and characteristics of conic sections.

In conclusion, analytic geometry provides a robust framework for understanding and analyzing geometric shapes and their properties. By translating geometric problems into algebraic equations, it enables the precise calculation of perimeters, areas, and volumes, as well as the exploration of the intricate properties of conic sections. This mathematical discipline is indispensable for engineers preparing for the FE Mechanical exam, as it lays the foundation for solving a wide range of engineering problems related to design, analysis, and optimization of mechanical systems.

Calculus: Differential and Integral Concepts

Calculus, a fundamental branch of mathematics, is pivotal in the field of engineering, providing tools for modeling and solving problems across various disciplines. Differential calculus focuses on the concept of the derivative, which represents the rate of change of a function with respect to a variable. Integral calculus, on the other hand, deals with the accumulation of quantities, leading to the concept of the integral. Both concepts are essential for understanding and predicting system behaviors in real-world engineering scenarios.

Differential Calculus begins with the definition of the derivative of a function $f(x)$ as the limit of the difference quotient as the interval approaches zero: $\lim_{h \to 0} \frac{f(x+h) - f(x)}{h}$. This concept is crucial for determining the slope of a tangent to a curve at any point, allowing engineers to analyze rates of change in physical systems. For instance, in mechanical engineering, the rate at which a component's velocity changes with time—its acceleration—can be determined using differential calculus.

Integral Calculus is introduced through the concept of the definite integral, defined as the limit of a sum: $\int_a^b f(x)dx$, which represents the area under the curve $f(x)$ from a to b. This is vital for calculating quantities such as the distance traveled by an object when given its velocity as a function of time. Furthermore, the Fundamental Theorem of Calculus links the concept of differentiation and integration, providing a powerful tool for solving problems involving accumulation of quantities and rates of change.

In engineering, **multiple variables** often influence systems, necessitating the extension of calculus to functions of several variables. **Partial derivatives** and **multiple integrals** are the tools used to handle such situations. The partial derivative of a function with respect to one variable, while keeping other variables constant, shows how a function changes as that specific variable changes. This is essential in fields like thermodynamics and fluid mechanics, where properties like temperature and pressure depend on multiple variables. Multiple integrals, on the other hand, allow for the calculation of volumes under surfaces defined by functions of two variables, crucial for designing and analyzing complex systems.

Real-world applications of calculus in mechanical engineering are vast. For example, in the analysis of dynamic systems, differential equations—equations involving derivatives—describe the relationship between forces acting on a system and its motion. Solving these equations enables engineers to predict system behavior under various conditions. In material science, understanding how material stress and strain relate through differential equations helps in designing safer and more efficient structures.

Calculus also plays a critical role in **optimization problems**, where engineers seek to determine the best solution among a set of possible options. Techniques involving the derivative, such as finding where the derivative equals zero, can identify maximum or minimum points of functions representing cost, efficiency, or other critical parameters.

In summary, calculus is an indispensable tool in the arsenal of an engineer, providing the means to model, analyze, and solve complex problems by understanding and manipulating rates of change and accumulations. Its applications span all branches of engineering, making a solid grasp of calculus principles essential for success in the FE Mechanical exam and beyond.

Ordinary Differential Equations

Ordinary Differential Equations (ODEs) are foundational to understanding the dynamics of mechanical systems and are pivotal in the analysis and design of engineering solutions. ODEs are equations that involve functions and their derivatives, representing rates of change. These equations are classified as either **homogeneous** or **nonhomogeneous**, depending on the presence of a term independent of the function in the equation.

Homogeneous ODEs are characterized by the absence of a term that is independent of the unknown function. They take the form $Ly = 0$, where L is a differential operator applied to the function y. The solutions to these equations are critical in understanding natural modes of vibration, thermal conduction, and fluid flow in engineering systems. The general approach to solving a homogeneous ODE involves finding the roots of the characteristic equation associated with the differential operator. For example, the second-order homogeneous ODE $ay'' + by' + cy = 0$ has a characteristic equation $ar^2 + br + c = 0$. The roots of this equation, real or complex, dictate the form of the solution, which could be exponential, oscillatory, or a combination of both.

Nonhomogeneous ODEs, on the other hand, include an additional term that does not depend on the unknown function, represented as $Ly = f(x)$, where $f(x)$ is a known function. The presence of $f(x)$ introduces external forces or inputs into the system, such as non-zero boundary conditions, forcing functions in mechanical vibrations, or heat sources in thermal conduction problems. The solution to a nonhomogeneous ODE is the sum of the homogeneous solution and a particular solution that accounts for the nonhomogeneous term. Finding the particular solution often involves methods such as undetermined coefficients or variation of parameters, depending on the form of $f(x)$.

The **Laplace transform** is a powerful tool for solving both homogeneous and nonhomogeneous ODEs, especially when dealing with complex boundary conditions or initial value problems. The Laplace transform converts differential equations in the time domain into algebraic equations in the s-domain, where they can be solved more straightforwardly. The transformed equation is then inverted back to the time domain to obtain the solution. This method is particularly useful in engineering for analyzing systems subject to step, impulse, or sinusoidal inputs, as it simplifies the process of finding solutions that satisfy given initial conditions.

For instance, consider the second-order nonhomogeneous ODE $y'' + 2y' + 5y = e^{-t}$ with initial conditions $y(0) = 0$ and $y'(0) = 0$. Applying the Laplace transform to both sides of the equation and using the initial conditions transforms the problem into an algebraic equation in terms of $Y(s)$, the Laplace transform of $y(t)$. Solving for $Y(s)$ and then applying the inverse Laplace transform yields the solution in the time domain, $y(t)$, which describes how the system responds over time to the external input e^{-t}.

Linear Algebra in Engineering

Linear Algebra serves as a cornerstone in the realm of engineering, providing essential tools for modeling, analyzing, and solving complex systems. At the heart of linear algebra are **matrix operations**, which include addition, subtraction, and, most critically, multiplication. These operations allow engineers to manipulate arrays of numbers or variables systematically, facilitating the representation of systems of linear equations, transformations, and more. For instance, the multiplication of a matrix by a vector can represent the transformation of a geometric object in space, a fundamental concept in computer graphics and simulation.

Determinants are scalar values that can be computed from a square matrix and offer profound insights into the properties of the matrix. A non-zero determinant indicates that the matrix is invertible, an essential condition for solving systems of linear equations. The determinant can also be interpreted in geometrical terms as a scaling factor for volume when a matrix is used to transform geometric shapes. In engineering, determinants are used in analysis and stability studies, such as determining the eigenvalues of a system, which are crucial for understanding the behavior of structures or mechanical vibrations.

Vector analysis extends the utility of linear algebra by focusing on the magnitude and direction of quantities. Vectors are fundamental in fields such as fluid dynamics, electromagnetism, and mechanics, where they represent forces, velocities, and other directional quantities. Operations such as dot product and cross product are vital in vector analysis, enabling the calculation of work done by a force or the torque on a rotating system, respectively.

The application of linear algebra in engineering problems is vast and varied. In **structural engineering**, matrix methods like the stiffness matrix allow for the analysis of complex structures by simplifying them into a system of linear equations. This approach is invaluable in designing buildings and bridges, ensuring they can withstand loads and stresses. **Electrical engineering** relies heavily on linear algebra for circuit analysis, where matrices represent the relationships between voltages, currents, and resistances in a circuit, allowing for the efficient calculation of circuit behavior.

In **control systems engineering**, linear algebra is used to model and analyze the behavior of systems over time. The state-space representation, a model comprising matrices for system states and inputs, is a powerful framework for designing controllers that ensure systems behave as desired. This is critical in applications ranging from automatic pilot avionics to industrial process control.

Furthermore, linear algebra's role in **data analysis and machine learning** cannot be overstated. Techniques such as singular value decomposition (SVD) and principal component analysis (PCA) rely on linear algebra to process and reduce the dimensionality of large datasets, enabling the extraction of meaningful patterns and the making of predictions. This has applications in fields ranging from bioinformatics to finance, where engineers and scientists use these methods to analyze complex data.

The depth and breadth of linear algebra's applications in engineering underscore its importance in the FE Mechanical exam and beyond. Mastery of matrix operations, determinants, and vector analysis equips aspiring engineers with the mathematical foundation necessary for tackling a wide array of engineering challenges, ensuring their solutions are not only effective but optimized for efficiency and innovation.

Numerical Methods and Error Analysis

Numerical methods serve as a cornerstone in engineering problem-solving, particularly when analytical solutions are intractable or non-existent. These methods rely on numerical approximation to model and solve complex equations, making them indispensable in the FE Mechanical exam preparation. **Approximation techniques** form the basis of numerical analysis, providing a means to estimate solutions with a known degree of accuracy. One fundamental approach is the use of **Taylor's series**, which represents functions as infinite sums of their derivatives at a single point. The Taylor series for a function $f(x)$ about a point $x = a$ is given by

$$f(x) = f(a) + f'(a)(x - a) + \frac{f''(a)}{2!}(x - a)^2 + \frac{f'''(a)}{3!}(x - a)^3 + \cdots$$

This expansion is particularly useful in engineering for approximating the values of functions near a known point, enabling the analysis of systems where exact solutions are cumbersome to obtain.

Error analysis is another critical aspect, providing insights into the precision and reliability of numerical approximations. It encompasses the study of **round-off errors**, arising from the finite precision of numerical representations, and **truncation errors**, resulting from the use of approximations instead of exact mathematical procedures. Understanding these errors is crucial for engineers to ensure that their numerical solutions are both accurate and practical for real-world applications.

Newton's method, also known as the Newton-Raphson method, is a powerful technique for finding successively better approximations to the roots (or zeroes) of a real-valued function. The method starts with an initial guess x_0 for a root of the function $f(x)$, and iteratively refines this guess using the formula:

$$x_{n+1} = x_n - \frac{f(x_n)}{f'(x_n)}$$

This process is repeated until a sufficiently accurate value is obtained. Newton's method is highly valued for its rapid convergence to a solution, making it a preferred tool in engineering for solving equations where direct solutions are not feasible.

The integration of these numerical methods into engineering practice is not merely academic; it is a pragmatic approach to tackling the complex, nonlinear problems frequently encountered in mechanical engineering. From designing components to analyzing systems, the ability to apply numerical methods like Taylor's series and Newton's method, coupled with a thorough understanding of error analysis, equips engineers with the skills to develop innovative solutions that are both efficient and robust. This proficiency is not only pivotal for excelling in the FE Mechanical exam but also forms the foundation for advanced engineering analysis and design in professional practice.

Algorithm and Logic Development

In the realm of engineering, the development of algorithms and logic is a critical skill that enables engineers to solve complex problems efficiently and effectively. This section delves into the fundamentals of algorithm development, focusing on the use of **flowcharts** and **pseudocode** as tools for structured problem-solving and fostering algorithmic thinking.

Flowcharts are visual representations of the steps involved in a process or system. They employ standardized symbols such as ovals, rectangles, and diamonds to denote start/end points, processing steps, and decision points, respectively. Flowcharts are instrumental in laying out the sequence of actions in a clear, logical manner, making it easier to understand and communicate the process. For instance, in designing a control system for a manufacturing process, a flowchart can visually map out the sequence of operations, from raw material input to final product output, including quality checks and decision-making points regarding material quality.

Pseudocode, on the other hand, is a method used to describe algorithms using the structural conventions of programming languages, but in a more natural language. It allows the designer to focus on the algorithm's logic without getting bogged down by the syntax of actual programming languages. Pseudocode is an essential step in algorithm development, bridging the gap between the problem-solving process and the actual coding. It outlines the algorithm in a way that is easily translated into any programming language, making it a versatile tool for engineers across disciplines. For example, an algorithm for a temperature control system might be outlined in pseudocode before being implemented in a specific programming language, ensuring the logic is sound and all scenarios are accounted for.

The process of developing an algorithm typically begins with defining the problem and understanding the requirements. This is followed by brainstorming and conceptualizing potential solutions, selecting the most viable solution, and then representing this solution in the form of a flowchart or pseudocode. This representation is then used to create a detailed algorithm that can be implemented in code.

Consider the problem of sorting a list of numbers in ascending order. The first step is to understand the sorting process and what the final sorted list should look like. Next, one might conceptualize different sorting methods, such as bubble sort, selection sort, or insertion sort. After selecting a method, a flowchart can be created to visualize the process step by step, such as comparing each number with the next, swapping them if they are in the wrong order, and repeating this process until the entire list is sorted. The same process can be outlined in pseudocode, specifying the steps in a format that closely resembles the structure of a programming language but is easier to read and understand.

Both flowcharts and pseudocode play a crucial role in algorithm development, serving as intermediate steps between problem identification and coding. They help in identifying logical errors or inefficiencies in the proposed solution early in the development process, saving time and resources in the long run. Moreover, they are invaluable tools for documentation and collaboration, allowing engineers to share and review algorithms with peers or integrate them into larger systems with ease.

Chapter 2: Probability and Statistics

Probability Distributions and Applications

Understanding probability distributions is fundamental for engineers to model uncertainty and make informed decisions under conditions of variability. Probability distributions can be broadly classified into discrete and continuous types, each with its specific applications and characteristics. **Discrete distributions** are used when the set of possible outcomes is countable. Among the most significant discrete distributions is the **Binomial Distribution**, which models the number of successes in a fixed number of independent Bernoulli trials, each with the same probability of success. The probability mass function (PMF) of a binomial distribution is given by $P(X = k) = \binom{n}{k} p^k (1-p)^{n-k}$, where n is the number of trials, k is the number of successes, p is the probability of success on a single trial, and $\binom{n}{k}$ is the binomial coefficient.

Another important discrete distribution is the **Poisson Distribution**, which is used for modeling the number of times an event occurs in a fixed interval of time or space. The PMF of a Poisson distribution is $P(X = k) = \dfrac{\lambda^k e^{-\lambda}}{k!}$, where λ is the average number of events in an interval, and k is the actual number of events observed. The Poisson distribution is particularly useful for modeling rare events in a large population over a short period.

Continuous distributions, on the other hand, are used when the set of possible outcomes is uncountable, typically involving measurements. The **Normal Distribution**, also known as the Gaussian distribution, is a cornerstone in the field of statistics and engineering due to its natural occurrence in many physical, biological, and social phenomena. The probability density function (PDF) of a normal distribution is $f(x) = \dfrac{1}{\sigma \sqrt{2\pi}} e^{-\frac{1}{2}\left(\frac{x-\mu}{\sigma}\right)^2}$, where μ is the mean, and σ is the standard deviation. The normal distribution is symmetric about the mean, and its shape is determined entirely by μ and σ.

The **Exponential Distribution** is another continuous distribution, often used to model the time between events in a Poisson process. The PDF of an exponential distribution is $f(x) = \lambda e^{-\lambda x}$ for $x \geq 0$, where λ is the rate parameter. This distribution is particularly useful for reliability engineering and failure analysis, as it can model the time until failure of components or systems.

Empirical distributions are derived from data without assuming any underlying theoretical distribution. They are particularly useful for modeling phenomena when the theoretical distribution is unknown or when working with small sample sizes. Empirical distributions can be represented using histograms or cumulative distribution functions (CDFs) constructed from observed data.

In engineering applications, understanding and applying the appropriate probability distribution allows for the modeling of uncertainties in processes and systems. For instance, the binomial distribution can model the reliability of a batch of components, predicting the number of failures. The normal distribution is widely used in quality control to determine the variability of manufactured products. The exponential distribution helps in assessing the lifespan of machinery and equipment, facilitating maintenance scheduling and reliability analysis.

Central Tendency and Dispersion Measures

Understanding measures of central tendencies and dispersions is crucial for interpreting data effectively, especially in the context of engineering where decision-making often relies on statistical analysis. The mean, mode, and standard deviation are foundational concepts in statistics that provide insights into the distribution and variability of a data set. Confidence intervals add another layer of analysis, offering a range within which we can expect the true population parameter to lie with a certain level of confidence.

The **mean** is the arithmetic average of a set of values, calculated by summing all the values and dividing by the number of values. It is represented by the formula $\bar{x} = \frac{\sum_{i=1}^{n} x_i}{n}$, where $\bar{x}$ is the sample mean, x_i represents each value in the sample, and n is the total number of values. The mean is sensitive to outliers, which can skew its value and may not accurately represent the central tendency of highly skewed distributions.

The **mode** refers to the most frequently occurring value in a data set. A data set may have one mode (unimodal), two modes (bimodal), or more modes (multimodal). The mode is particularly useful for categorical data where we wish to identify the most common category.

Standard deviation measures the dispersion or variability of a data set, indicating how spread out the values are from the mean. A low standard deviation means that the data points are close to the mean, while a high standard deviation indicates that the data points are spread out over a wider range of values. The formula for the sample standard deviation is $s = \sqrt{\frac{\sum_{i=1}^{n}(x_i - \bar{x})^2}{n-1}}$

, where s is the sample standard deviation, x_i represents each value, $\bar{x}$ is the sample mean, and n is the number of values.

Confidence intervals provide a range of values within which we can expect the true population parameter (mean, proportion) to lie with a certain level of confidence, typically expressed as 90%, 95%, or 99%. The confidence interval for the mean is calculated as $\bar{x} \pm z\dfrac{s}{\sqrt{n}}$, where $\bar{x}$ is the sample mean, z is the z-score corresponding to the desired confidence level, s is the standard deviation, and n is the sample size. Confidence intervals are essential for understanding the precision of an estimate and the reliability of the data.

These statistical measures are applied to analyze and interpret data from experiments, quality control, and research studies, among other applications. For instance, understanding the mean and standard deviation of material properties can inform design decisions, while confidence intervals can guide engineers in assessing the reliability of these properties under different conditions.

Expected Value

The concept of **expected value** is a cornerstone in the field of probability and statistics, particularly when it comes to making informed decisions under conditions of uncertainty. It represents the average outcome one can anticipate from a random event when the process is repeated a large number of times. The expected value is essentially a weighted average of all possible outcomes, with the weights being the probabilities of each outcome occurring. This concept is not only pivotal in various branches of engineering but also in economics, finance, and decision theory, where it aids in the evaluation of the potential benefits and risks associated with different decisions.

To calculate the expected value, one multiplies each possible outcome by the probability of that outcome, and then sums all these products. The formula for expected value $E(X)$ of a discrete random variable X with possible values $x_1, x_2, ..., x_n$ and corresponding probabilities $P(x_1), P(x_2), ..., P(x_n)$ is given by:

$$E(X) = \sum_{i=1}^{n} x_i P(x_i)$$

Where x_i represents a possible value of X, and $P(x_i)$ is the probability of x_i occurring.

For continuous random variables, the expected value is calculated using an integral over the entire range of possible outcomes, which represents the probability density function of the variable. The formula for the expected value of a continuous random variable X with probability density function $f(x)$ is:

$$E(X) = \int_{-\infty}^{\infty} x f(x) dx$$

Understanding the expected value is crucial for engineers as it provides a method to quantify the average outcome of processes that involve randomness. For instance, in the context of quality control, an engineer might use the expected value to calculate the average defect rate in a batch of materials. Similarly, in financial engineering, the expected value can help in assessing the average return or risk associated with different investment options.

Moreover, the concept of expected value is instrumental in optimizing systems and processes. Engineers often face scenarios where they must choose between multiple design options, each with its own set of probabilistic outcomes. By calculating the expected value of each option, engineers can objectively evaluate which design maximizes efficiency, minimizes cost, or achieves the best balance between competing objectives.

In practice, the calculation of expected value requires a thorough understanding of the system or process being analyzed, including the identification of all possible outcomes and their associated probabilities. This often involves statistical analysis and modeling to accurately estimate probabilities, especially in complex systems where outcomes are influenced by a multitude of factors.

While the concept of expected value is mathematically straightforward, its application can be complex, particularly in engineering problems where outcomes and probabilities are not always clear-cut. Nonetheless, by applying this concept, engineers can make more informed decisions, leading to better outcomes in projects and processes characterized by uncertainty.

Regression and Curve Fitting

Regression and curve fitting are fundamental statistical tools used to model and analyze the relationship between variables. Linear regression, one of the most basic forms of regression, involves finding the best-fitting straight line through a set of data points. This line is described by the equation $y = mx + b$, where y represents the dependent variable, x the independent variable, m the slope of the line, and b the y-intercept. The goal of linear regression is to determine the values of m and b that minimize the difference between the observed values and the values predicted by the model.

Multiple regression extends the concept of linear regression by incorporating two or more independent variables. The equation for a multiple regression model can be represented as $y = b_0 + b_1 x_1 + b_2 x_2 + \ldots + b_n x_n$, where y is the dependent variable, $x_1, x_2, \ldots, x_n$ are the independent variables, and $b_0, b_1, \ldots, b_n$ are the coefficients that represent the relationship between each independent variable and the dependent variable. Multiple regression allows for the analysis of the combined effect of several factors on a single outcome, making it a powerful tool for engineering applications where multiple variables influence a response.

Curve fitting involves finding a curve that best fits a series of data points. Unlike linear regression, which is limited to linear models, curve fitting can involve linear or nonlinear models. The choice of model depends on the nature of the relationship between the variables. Nonlinear models can take various forms, such as polynomial equations $y = ax^2 + bx + c$ for quadratic models or more complex equations for exponential, logarithmic, or sinusoidal fits. The objective is to select a model that closely approximates the underlying function that generated the data.

The goodness of fit of a model is a measure of how well it describes the observed data. One common metric for assessing goodness of fit is the coefficient of determination, denoted as R^2, which quantifies the proportion of the variance in the dependent variable that is predictable from the independent variables. An R^2 value of 1 indicates a perfect fit, while a value of 0 indicates that the model does not explain any of the variance in the data.

The least squares method is a standard approach for estimating the parameters of a regression model. It involves minimizing the sum of the squares of the differences between the observed values and the values predicted by the model. For linear regression, this leads to a set of normal equations that can be solved analytically to find the optimal values of the coefficients. For nonlinear models, numerical methods are typically used to find the parameters that minimize the sum of squared errors.

In engineering, regression and curve fitting are used for tasks such as predicting material properties, analyzing stress-strain data, and modeling the behavior of dynamic systems. By understanding the statistical principles behind these methods, engineers can develop models that accurately represent complex phenomena, enabling informed decision-making and optimization of processes and designs.

Chapter 3: Ethics and Professional Practice

Codes of Ethics in Engineering

The **NCEES Model Law** and the **standards from professional societies** form the cornerstone of ethical practice in engineering. These guidelines are designed to uphold the integrity, honor, and dignity of the engineering profession, ensuring that engineers act in a manner that is both legally compliant and ethically sound. The NCEES Model Law, in particular, provides a framework that regulates the qualifications for licensure, standards for practice, and disciplinary measures for those who fail to adhere to its principles. It emphasizes the paramount responsibility of engineers to protect the public health, safety, and welfare above all other considerations.

Professional societies such as the American Society of Mechanical Engineers (ASME), the Institute of Electrical and Electronics Engineers (IEEE), and the American Society of Civil Engineers (ASCE) have developed their own codes of ethics. These codes serve as a detailed guide for professional conduct, addressing issues such as conflict of interest, bribery, and corruption. They also highlight the importance of continuing education, professional development, and the advancement of the engineering profession through innovation and research.

Ethical and legal responsibilities in engineering are not merely about adhering to the letter of the law or the specific wording of ethical codes. They are about fostering a culture of integrity and accountability. Engineers are often faced with decisions that have significant societal impacts, from the design of safe transportation systems to the development of sustainable energy solutions. In making these decisions, engineers must consider the long-term welfare of the community and the environment, even when faced with pressures that may push them towards more expedient or profitable outcomes.

The ethical guidelines emphasize the importance of **confidentiality** in professional relationships, protecting the intellectual property rights of others, and avoiding deceptive acts that might compromise the integrity of engineering decisions or the safety of the public. Engineers are encouraged to report any unsafe, illegal, or unethical conduct they observe, even when doing so may be uncomfortable or unpopular.

In the context of the FE Mechanical exam, understanding these ethical and legal responsibilities is crucial. Not only do they form part of the examination content, but they also prepare candidates for the professional challenges they will face in their careers. As engineers progress towards obtaining their Professional Engineer (PE) license, their adherence to these ethical standards becomes even more critical. The PE license is not just a certification of technical competence; it is a testament to the engineer's commitment to upholding the highest standards of professional conduct.

The NCEES Model Law and the codes of ethics from professional societies provide a comprehensive framework for ethical engineering practice. They remind engineers of their obligation to society, the environment, and the profession itself. By adhering to these principles, engineers not only ensure their own success but also contribute to the advancement and integrity of the engineering profession as a whole.

Public Health, Safety, and Welfare

Engineers play a pivotal role in safeguarding public health, safety, and welfare, a responsibility that transcends the mere application of technical skills and knowledge. This duty is deeply embedded in the ethical fabric of the profession, demanding a commitment to protect the well-being of the community and the environment. The essence of this responsibility lies in the recognition that engineering decisions and actions have far-reaching consequences, affecting not only the immediate users of a product or service but also the broader public and future generations.

The ethical practice in engineering that prioritizes public health, safety, and welfare is guided by several key principles. First and foremost is the principle of non-maleficence, which obligates engineers to avoid harm to others. This principle is not limited to preventing physical harm but also encompasses preventing potential risks that could jeopardize the health and safety of the public. Engineers must, therefore, conduct thorough risk assessments and safety analyses for every project, considering both the short-term and long-term impacts of their work. This involves identifying potential hazards, assessing the likelihood and severity of harm, and implementing measures to mitigate risks to acceptable levels.

Another critical principle is beneficence, which compels engineers to contribute to the well-being of others. This principle encourages engineers to use their skills and knowledge to improve public health, safety, and welfare actively. For instance, engineers can design safer transportation systems, develop technologies for clean water and sanitation, and create more efficient and less polluting energy systems. By focusing on projects that offer significant benefits to society, engineers can make substantial contributions to public welfare.

The principle of justice requires engineers to ensure that the benefits and burdens of their work are distributed fairly among all members of society. This includes paying special attention to the needs of the most vulnerable populations and ensuring that engineering projects do not disproportionately impact certain groups negatively. Engineers must strive for equitable access to the benefits of technology and work to prevent environmental injustice, where marginalized communities bear the brunt of pollution and other adverse effects of industrial activities.

Transparency and honesty are also vital to ethical engineering practice. Engineers must communicate openly about the risks and benefits of their projects, ensuring that stakeholders, including the public, are fully informed and able to make educated decisions. This includes disclosing potential conflicts of interest and avoiding any actions that could deceive or mislead others about the safety, performance, or impact of engineering projects.

In fulfilling their ethical obligations to safeguard public health, safety, and welfare, engineers must adhere to the highest standards of professional conduct. This includes maintaining competence through continual learning, adhering to applicable laws and regulations, and following the codes of ethics established by professional societies. By upholding these ethical principles, engineers not only protect the public but also enhance the trust and respect that society places in the engineering profession.

The role of engineers in safeguarding public health, safety, and welfare is a testament to the profound impact that engineering has on society. It underscores the importance of ethical practice in ensuring that technological advancements serve the common good, enhancing the quality of life for all. As engineers navigate the complexities of modern engineering challenges, their commitment to ethical principles will remain essential in guiding their efforts to protect and improve the well-being of the public.

Intellectual Property in Engineering

In the realm of engineering, the safeguarding of intellectual property (IP) is paramount, not only as a legal obligation but also as an ethical practice that fosters innovation, creativity, and fair competition. Intellectual property rights are designed to protect the creations of the mind, including inventions, literary and artistic works, designs, symbols, names, and images used in commerce. For engineers, understanding and respecting these rights are crucial in navigating the professional landscape, ensuring that their contributions and the contributions of others are appropriately recognized and protected.

Copyrights are a form of protection provided to the authors of "original works of authorship," including literary, dramatic, musical, architectural, cartographic, choreographic, pictorial, graphic, sculptural, and audiovisual creations. For engineers, this can encompass software code, technical reports, design drawings, and manuals. Copyright does not protect ideas, only the expression of those ideas. It is automatically secured upon creation and fixed in a tangible form that is perceptible either directly or with the aid of a machine or device.

Patents grant inventors exclusive rights to their inventions, allowing them to exclude others from making, using, selling, or importing the invention for a limited period, typically 20 years from the filing date of the patent application. In engineering, patents are crucial for protecting innovative solutions, processes, machines, and chemical compositions. Obtaining a patent requires a detailed application that demonstrates the invention's novelty, non-obviousness, and utility. This process is complex and often requires the expertise of a patent attorney or agent.

Trademarks protect symbols, names, and slogans used to identify goods or services. The primary purpose of a trademark is to prevent confusion in the marketplace, ensuring that consumers can distinguish between products and services. For engineering firms, trademarks can protect brand names, product names, logos, and slogans. Unlike patents and copyrights, trademark rights can last indefinitely, provided the mark remains in use and retains its distinctive character.

Trade secrets encompass formulas, practices, processes, designs, instruments, patterns, or compilations of information that are not generally known or reasonably ascertainable, by which a business can obtain an economic advantage over competitors or customers. In engineering, trade secrets might include proprietary manufacturing processes, algorithms, or client lists. Protection of trade secrets does not require registration; however, it necessitates reasonable steps by the owner to keep the information secret, such as non-disclosure agreements (NDAs).

Respecting intellectual property rights involves several best practices. Engineers should always seek permission before using copyrighted materials, ensure that patent applications are filed before public disclosure of an invention, use trademarks correctly to maintain their distinctiveness and protectiveness, and implement measures to safeguard trade secrets, including confidentiality agreements and secure information systems.

Violations of intellectual property rights can lead to legal disputes, financial penalties, and damage to professional reputation. Therefore, engineers must be vigilant in their creation, use, and management of intellectual property, adhering to both the letter and spirit of IP laws and ethical guidelines. This not only protects their work and the work of others but also promotes a culture of integrity and respect within the engineering community.

Societal Impact of Engineering

Engineers play a pivotal role in shaping society through their contributions to the economy, sustainability, life-cycle management, and environmental stewardship. Their decisions have far-reaching implications that extend beyond the immediate scope of their projects, influencing the well-being of communities and the planet. The integration of ethical considerations into engineering practice is essential for fostering a sustainable future and ensuring that technological advancements contribute positively to society.

The economy benefits significantly from engineering innovations, which drive growth, create jobs, and improve quality of life. Engineers are at the forefront of developing new technologies and infrastructure that boost productivity and efficiency. However, it is crucial that these economic advancements do not come at the expense of environmental degradation or social inequality. Sustainable engineering practices aim to balance economic growth with environmental stewardship and social equity, ensuring that development meets the needs of the present without compromising the ability of future generations to meet their own needs.

Sustainability in engineering encompasses the design, construction, and operation of systems that use resources efficiently while minimizing environmental impact. This involves considering the entire life cycle of products and projects, from material extraction and manufacturing to use and end-of-life disposal or recycling. Engineers must evaluate the environmental implications of their designs, including energy consumption, waste generation, and greenhouse gas emissions. By adopting principles of sustainable design, such as using renewable energy sources, selecting eco-friendly materials, and implementing energy-efficient technologies, engineers can contribute to the preservation of natural resources and the mitigation of climate change.

Life-cycle assessment (LCA) is a systematic approach used by engineers to assess the environmental aspects and potential impacts associated with a product, process, or service throughout its life cycle. LCA helps identify opportunities to improve the environmental performance of products at various stages of their life cycle, from raw material extraction through manufacturing, use, and disposal. By understanding the life-cycle impacts of their designs, engineers can make informed decisions that reduce negative environmental effects and enhance the sustainability of their projects.

The environment is directly affected by engineering activities, and responsible engineering practices are essential for protecting ecosystems and ensuring a healthy planet. Engineers must consider the potential environmental impacts of their projects, including pollution, habitat destruction, and resource depletion. Environmental impact assessments (EIA) are a critical tool for evaluating the potential effects of engineering projects on the environment and identifying measures to mitigate negative impacts. Engineers have a responsibility to incorporate environmental considerations into their designs, promoting conservation efforts, pollution prevention, and the restoration of natural habitats.

In conclusion, engineers have a significant influence on society through their contributions to the economy, sustainability, life-cycle management, and environmental protection. Ethical and responsible decision-making is crucial for ensuring that engineering projects benefit society as a whole, fostering economic development while preserving the environment for future generations. By integrating principles of sustainability, life-cycle thinking, and environmental stewardship into their practices, engineers can lead the way in creating a more sustainable and equitable world.

Chapter 4: Engineering Economics

Time Value of Money Concepts

The concept of the **Time Value of Money (TVM)** is foundational in understanding how money's worth changes over time. This principle posits that a dollar today holds more value than a dollar in the future due to its potential earning capacity. This core concept underpins various financial calculations and decisions, including present worth, future worth, rate of return, and annuities, which are crucial for engineers to master when making economic evaluations of projects over time.

Present Worth (PW) or Present Value (PV) is the current value of a future amount of money or stream of cash flows given a specified rate of return. The formula for calculating present worth is $PV = \dfrac{FV}{(1+r)^n}$, where FV is the future value, r is the interest rate, and n is the number of periods. This calculation helps in assessing the value today of a sum of money to be received in the future, enabling comparisons between investments of different time horizons.

Future Worth (FW) or Future Value (FV) is the amount of money an investment will grow to over a period of time at a given interest rate. The future worth can be calculated using the formula $FV = PV \times (1+r)^n$, where PV is the present value, r is the interest rate, and n is the number of periods. This formula is pivotal for engineers in projecting the future value of current investments or costs.

Rate of Return (ROR) is the percentage of increase or decrease in an investment over a period of time, compared to the initial investment. It is a critical measure for evaluating the profitability of potential projects. The rate of return can be calculated using the formula $ROR = \dfrac{FV - PV}{PV} \times 100\%$, where FV is the future value of the investment, and PV is the present value of the investment.

Annuities are a series of equal payments made at regular intervals over a period. Annuities can be ordinary (payments at the end of each period) or due (payments at the beginning of each period). The formula for the present value of an ordinary annuity is $PV = P \times \dfrac{1 - (1+r)^{-n}}{r}$, where P is the payment amount, r is the interest rate per period, and n is the number of periods. Understanding annuities allows engineers to evaluate the worth of uniform series of payments or receipts over time, which is common in financing and cost analysis of engineering projects.

In applying these concepts, engineers can make informed decisions regarding the economic viability of projects, investments, and other financial considerations in their professional practice. The ability to calculate and interpret present worth, future worth, rate of return, and the value of annuities is indispensable for conducting thorough economic analyses and ensuring the financial success of engineering projects.

Cost Types and Breakdowns

Understanding the various types of costs is crucial for engineers to conduct detailed financial analysis and make informed decisions regarding project management, budgeting, and economic evaluation. **Fixed costs** are expenses that do not change with the level of production or service activity within a certain range. These costs are constant regardless of the business's operational output, such as rent, salaries, and insurance premiums. They are essential for budgeting and financial planning as they are predictable over the short term.

Variable costs, on the other hand, fluctuate with the level of production or service activity. These costs include raw materials, direct labor, and any other expenses directly tied to the production volume. Understanding variable costs is vital for break-even analysis and operational planning since they directly impact the marginal cost of production and, consequently, pricing strategies.

Incremental costs, also known as marginal costs, refer to the additional costs incurred when increasing the level of production or service by one unit. This concept is pivotal in decision-making processes, especially when evaluating the financial implications of expanding production or introducing a new product line. Incremental cost analysis helps in determining the optimal production level and in making cost-effective decisions regarding resource allocation.

Average costs represent the total costs (fixed plus variable) divided by the number of units produced. This metric is beneficial for pricing decisions, allowing businesses to determine the minimum selling price needed to cover all costs and achieve a desired profit margin. It also aids in performance evaluation by comparing the average costs over different periods or among different projects.

Sunk costs are expenditures that have already been incurred and cannot be recovered. These costs should not influence future business decisions since they remain constant regardless of the outcome of future events. Understanding that sunk costs are irrelevant to future decisions is crucial for economic rationality, preventing good money from being thrown after bad in an attempt to justify past investment decisions.

Each of these cost types plays a significant role in the financial analysis and economic evaluation of engineering projects. By accurately identifying and categorizing costs, engineers can perform more precise cost-benefit analyses, improve budgeting accuracy, and enhance the overall financial management of projects. This knowledge empowers engineers to optimize resource allocation, minimize waste, and maximize profitability, thereby contributing to the financial success and sustainability of engineering endeavors.

Economic Analyses

Economic analyses are pivotal in determining the feasibility and sustainability of engineering projects. These analyses encompass a variety of methods, each tailored to provide insights into different aspects of project economics. **Cost-benefit analysis (CBA)** is a systematic approach to estimating the strengths and weaknesses of alternatives. It is used to determine options that provide the best approach to achieve benefits while preserving savings. The formula for CBA is $\text{Net Present Value (NPV)} = \sum \frac{B_t - C_t}{(1+r)^t}$, where B_t and C_t are the benefits and costs at time t, respectively, and r is the discount rate. This analysis helps in comparing the total expected cost of each option against the total expected benefits, to see whether the benefits outweigh the costs, and by how much.

Break-even analysis identifies the point at which the costs of producing a product equal the revenue made from selling the product. This analysis is crucial for understanding the minimum production necessary to cover costs, using the formula $\text{Break-Even Point (units)} = \frac{FixedCosts}{PriceperUnit - VariableCostperUnit}$. It aids in financial planning by determining the level of sales needed to cover the total fixed and variable expenses.

Minimum cost analysis focuses on finding the least cost option among different alternatives without compromising the quality and efficiency of the project. This involves detailed cost estimation and comparison of various project elements to identify the most cost-effective solution.

Overhead analysis deals with the ongoing expenses of operating a business or a project. These are the costs not directly tied to the creation of a product or service but necessary for the business's overall functionality, such as utilities, rent, and administrative salaries. Understanding overhead costs is essential for budgeting, pricing, and profitability analysis.

Life-cycle analysis (LCA) evaluates the total environmental impact of a product or service throughout its lifespan, from raw material extraction through materials processing, manufacture, distribution, use, repair and maintenance, and disposal or recycling. In economic terms, LCA helps in understanding the cradle-to-grave costs, ensuring that the project or product is not only economically viable in the short term but also sustainable in the long term. This involves calculating the cumulative economic costs associated with a product over its entire life cycle, which can influence design, material selection, and process decisions to minimize costs and environmental impact.

The various analyses are essential for the economic assessment of engineering projects. Utilizing these methods enables engineers to make decisions that consider cost, quality, and sustainability, which contributes to the ongoing success and feasibility of their projects. These economic instruments facilitate a structured evaluation of the financial components of projects, establishing a robust basis for strategic planning and decision-making.

Chapter 5: Electricity and Magnetism

Electrical Fundamentals

Understanding electrical fundamentals is crucial for grasping the behavior of circuits and the principles of electricity and magnetism. **Charge** is the fundamental property of matter that exhibits electrostatic attraction or repulsion in the presence of other matter. It is quantified in coulombs (C), where the electron carries a charge of approximately -1.602×10^{-19} C. **Current** (I), measured in amperes (A), represents the flow of electric charge per unit time, moving from the positive to the negative terminal in a circuit. Ohm's Law, $V = IR$, where V is voltage in volts (V), I is current in amperes (A), and R is resistance in ohms (Ω), describes how these quantities are interrelated. **Voltage**, or electric potential difference, is the work needed per unit charge to move a charge between two points in a circuit. **Resistance** is the opposition to the flow of current, influenced by materials' properties, length, and cross-sectional area.

Power (P), measured in watts (W), is the rate at which electrical energy is transferred by an electric circuit. The formula $P = IV$ combines the concepts of voltage and current to calculate power, illustrating how much work is done or energy is converted per unit time. **Energy** (E), measured in joules (J), is the capacity to do work. In electrical systems, it's often calculated over time as $E = P \times t$, where t is time in seconds (s), providing a measure of the total work done or energy used.

Magnetic flux (Φ), measured in webers (Wb), is a measure of the quantity of magnetism, considering the strength and the extent of a magnetic field. Faraday's Law of Electromagnetic Induction, $\epsilon = -N \frac{d\Phi}{dt}$, where ϵ is the induced voltage, N is the number of turns in a coil, and $\frac{d\Phi}{dt}$ is the rate of change of magnetic flux, highlights the relationship between changing magnetic fields and induced electrical currents.

DC Circuit Analysis

Direct current (DC) circuits are fundamental to understanding electrical engineering and are characterized by the flow of electric charge in one direction. Analyzing these circuits involves applying **Kirchhoff's laws** and **Ohm's law**, as well as understanding series and parallel configurations. **Kirchhoff's Current Law (KCL)** states that the total current entering a junction equals the total current leaving the junction. This principle is crucial when analyzing complex circuits, ensuring that the conservation of charge is maintained. **Kirchhoff's Voltage Law (KVL)**, on the other hand, asserts that the sum of all electrical potential differences around any closed network is zero. This law is instrumental in calculating voltage drops across components in a loop, facilitating the analysis of energy conservation within the circuit.

Ohm's law is another cornerstone of DC circuit analysis, expressed as $V = IR$, where V is the voltage across the resistor, I is the current flowing through the resistor, and R is the resistance. This simple yet powerful relationship allows engineers to calculate unknown values in a circuit when two of the three variables are known, providing a foundational tool for electrical analysis and design.

When it comes to circuit configurations, **series** and **parallel** arrangements offer different behaviors in terms of voltage, current, and resistance. In a series circuit, components are connected end-to-end, so the same current flows through each component, but the voltage across each component can vary. The total resistance of a series circuit is the sum of the individual resistances, as given by $R_{total} = R_1 + R_2 + \ldots + R_n$. This configuration is often used when a uniform current is required through several components.

In contrast, a **parallel** circuit has components connected across the same voltage source, providing multiple paths for current flow. The voltage across each component in a parallel circuit is the same, but the total current is the sum of the currents through each parallel path. The total resistance in a parallel circuit is less than the smallest individual resistance and can be calculated using the formula $\frac{1}{R_{total}} = \frac{1}{R_1} + \frac{1}{R_2} + \ldots + \frac{1}{R_n}$. Parallel configurations are advantageous when components require the same voltage but may draw different amounts of current.

Grasping these principles and configurations is crucial for analyzing and designing DC circuits, forming the basis for more intricate electrical and electronic systems encountered in engineering. Utilizing Kirchhoff's laws, Ohm's law, and comprehending series and parallel circuits enables engineers to forecast and adjust the behavior of electrical systems to fulfill specific requirements, which is a vital skill for passing the FE Mechanical exam and achieving success in the engineering field.

AC Circuit Analysis

Alternating current (AC) circuit analysis is pivotal for understanding the dynamic behavior of electrical systems, especially those involving resistors, capacitors, and inductors. AC circuits are characterized by the periodic variation of the current and voltage, typically in a sinusoidal manner. This variation introduces complexities not present in direct current (DC) circuits, such as phase differences between voltage and current. To analyze these circuits, it is essential to employ phasor representation and complex impedance.

Phasor representation simplifies the analysis of AC circuits by converting sinusoidally varying voltages and currents into complex numbers, represented graphically as rotating vectors, or phasors. This method leverages Euler's formula, $e^{j\theta} = \cos\theta + j\sin\theta$, to express sinusoidal functions in terms of exponential functions with imaginary exponents, where j is the imaginary unit. For instance, a voltage $v(t) = V_m \cos(\omega t + \phi)$ can be represented as a phasor $V = V_m e^{j\phi}$, where V_m is the magnitude, ω is the angular frequency, and ϕ is the phase angle.

Complex impedance (Z), a fundamental concept in AC circuit analysis, extends the idea of resistance to capacitors and inductors, incorporating both magnitude and phase. The impedance of a resistor (R), capacitor (C), and inductor (L) are given by $Z_R = R$, $Z_C = \dfrac{1}{j\omega C}$, and $Z_L = j\omega L$, respectively. Here, $\omega = 2\pi f$, with f being the frequency of the AC supply. These expressions allow the analysis of circuits using Ohm's law ($V = IZ$) in the complex domain, facilitating the calculation of current and voltage across various components.

When analyzing circuits with resistors, capacitors, and inductors, it is crucial to consider the **frequency-dependent behavior** of capacitive and inductive reactances, $X_C = \dfrac{1}{\omega C}$ and $X_L = \omega L$. Capacitive reactance decreases with increasing frequency, while inductive reactance increases, affecting the overall impedance of the circuit and, consequently, the current and voltage distribution.

Kirchhoff's laws remain applicable in the analysis of AC circuits, with the caveat that they must be applied to phasors or complex numbers rather than real-valued instantaneous currents and voltages. **Kirchhoff's Current Law (KCL)** states that the algebraic sum of complex currents entering and leaving a junction equals zero, and **Kirchhoff's Voltage Law (KVL)** asserts that the algebraic sum of complex voltages around any closed loop equals zero.

The **power factor**, a measure of the phase difference between the voltage and current, is another critical aspect of AC circuit analysis. It is defined as the cosine of the phase angle ($\cos\phi$) between the voltage and current phasors. A power factor of 1 (or 100%) indicates that all the power is being effectively used, while a lower power factor signifies inefficiencies due to the phase difference, leading to wasted power in the form of reactive power.

Motors and Generators Basics

Electric motors and generators are fundamental components in the field of electrical engineering, embodying the principles of electromagnetic induction to convert energy from one form to another. At their core, both devices exploit the interaction between magnetic fields and electric currents to perform energy conversion. However, their roles are inverse; motors convert electrical energy into mechanical energy, while generators do the opposite, transforming mechanical energy into electrical energy.

The operation of electric motors is based on the Lorentz force principle, where a current-carrying conductor placed in a magnetic field experiences a force perpendicular to both the current and the field. This force is harnessed in motors to create rotational or linear motion. The basic construction of a motor involves a rotor (the moving part) and a stator (the stationary part), with one of these components generating a magnetic field that interacts with the current in the other, producing motion. The efficiency of an electric motor, often defined as the ratio of mechanical power output to electrical power input, is a critical parameter that influences the choice of a motor for a specific application. Various types of motors exist, including direct current (DC) motors, alternating current (AC) motors (such as synchronous and induction motors), and specialized types like stepper and servo motors, each suited to different applications based on characteristics like torque, speed control, and power consumption.

Generators, conversely, operate on the principle of electromagnetic induction discovered by Faraday, which states that a voltage is induced in a conductor when it is exposed to a changing magnetic field. This principle is exploited in generators to produce electrical voltage (and, consequently, current) by rotating a coil within a magnetic field or vice versa. The key components of a generator are similar to those of a motor, including a rotor and a stator, but function in reverse to convert mechanical energy into electrical energy. The mechanical energy required to turn the rotor can come from various sources, including steam turbines, water turbines, internal combustion engines, and even hand cranks. Generators are categorized into several types, such as AC generators (alternators) and DC generators (dynamos), with alternators being the most common type used in power generation due to their efficiency and the ease of transforming AC voltages.

The applications of motors and generators are vast and varied, encompassing almost every aspect of modern life. Motors are the workhorses behind most machinery, powering everything from household appliances to industrial equipment. Generators, on the other hand, are critical for power generation, providing electricity for homes, businesses, and even remote locations not connected to the grid. In renewable energy systems, such as wind turbines and hydroelectric plants, generators convert mechanical energy from wind or water flow into electrical energy, contributing to the global energy mix.

Grasping the principles and functioning of motors and generators is vital for engineers, as it plays a significant role in the design and optimization of these devices, as well as their integration into broader systems. The selection of various types of motors and generators, along with their sizing and incorporation into electrical and mechanical systems, relies on a comprehensive knowledge of their features, efficiencies, and appropriate applications. This expertise is crucial for succeeding in the FE Mechanical exam and for achieving success in the engineering field, where the capability to apply theoretical concepts to real-world situations is highly valuable.

Chapter 6: Statics

Resultants of Force Systems

In the analysis of static systems, understanding how to determine the resultant force from multiple forces acting on a system is crucial. This process involves both graphical and analytical methods to find a single force that represents the combined effect of all the forces acting on the body. The resultant force is a vector quantity, which means it has both magnitude and direction. To calculate this, one must consider both the scalar components of the forces in terms of their magnitudes and the angles at which they act.

Graphical Methods: The graphical approach, such as the parallelogram law, uses a scale drawing to determine the resultant force. By drawing the vectors to scale, in the direction they act, and completing the parallelogram, the diagonal of the parallelogram gives the magnitude and direction of the resultant force. This method, while intuitive, is less precise than analytical methods and is used for preliminary understanding or when precision is not critical.

Analytical Methods: The analytical approach is more precise and involves breaking down each force into its horizontal (F_x) and vertical (F_y) components using trigonometry. For a force F acting at an angle θ from the horizontal, the components can be calculated as $F_x = F\cos(\theta)$ and $F_y = F\sin(\theta)$. Once all forces are broken down into their components, the resultant force in each direction is found by summing up the respective components:

$$R_x = \sum F_x$$

$$R_y = \sum F_y$$

The magnitude of the resultant force (R) can then be determined using the Pythagorean theorem:

$$R = \sqrt{R_x^2 + R_y^2}$$

And the direction of the resultant force, θ_R, relative to the horizontal can be found using the tangent function:

$$\theta_R = \tan^{-1}\left(\frac{R_y}{R_x}\right)$$

This analytical method is not only precise but also allows for the solution of more complex problems where forces are not limited to two dimensions. In three-dimensional problems, the forces are broken down into their components along the x, y, and z axes, and the process is analogous to the two-dimensional case but includes the summation of the z-components (F_z) and the calculation of the resultant using all three components.

Understanding the principles of equilibrium is also essential when determining the resultant of force systems. A body is in equilibrium when the sum of all forces and the sum of all moments acting on it are zero. This principle is used to solve for unknown forces when the resultant force is known to be zero, a common scenario in many engineering problems.

Concurrent Force Systems

In the realm of statics, understanding concurrent force systems is pivotal for engineers aiming to analyze and design structures and mechanical systems efficiently. Concurrent forces are a collection of forces that act through a common point, albeit in various directions. The analysis of these forces is crucial for determining the resultant force, which is the single force that has the same effect on the body as the original system of forces. The equilibrium of a particle subjected to concurrent forces is a fundamental concept, as it forms the basis for more complex structural analyses.

The first step in analyzing concurrent force systems is to resolve each force into its components. Typically, this involves decomposing the forces into their x and y components, using the sine and cosine functions. For a force F making an angle θ with the horizontal, the horizontal component F_x can be calculated as $F_x = F\cos(\theta)$, and the vertical component F_y as $F_y = F\sin(\theta)$. This decomposition allows for a more straightforward addition of forces since forces in the same direction can be algebraically added.

Once the components of all forces are determined, the next step is to find the sum of the horizontal components $\sum F_x$ and the sum of the vertical components $\sum F_y$. These sums provide the components of the resultant force R, where $R_x = \sum F_x$ and $R_y = \sum F_y$. The magnitude of the resultant force can then be found using the Pythagorean theorem, given by $R = \sqrt{R_x^2 + R_y^2}$, and its direction relative to the horizontal is given by $\theta_R = \tan^{-1}\left(\frac{R_y}{R_x}\right)$.

In engineering practice, the ability to accurately analyze concurrent force systems is essential for ensuring the stability and integrity of structures and mechanical components. For instance, in the design of a truss structure, the joints are often modeled as particles where several members meet, exerting forces in different directions. Correctly determining the resultant force at each joint is critical for assessing whether the structure can withstand the applied loads without exceeding the strength limits of its components.

Moreover, the principles of concurrent force systems extend beyond static structures to include dynamic systems where forces may vary over time but still converge at a point. Understanding the instantaneous equilibrium of such systems is vital for the design and analysis of machinery and mechanical devices that undergo rapid motion changes.

The analysis of concurrent force systems is fundamental to the field of statics, allowing engineers to predict how structures and mechanical systems will respond to different load conditions. Proficiency in this area not only supports success on the FE Mechanical exam but also provides essential knowledge for future engineers in their professional careers. By systematically breaking down forces, performing algebraic addition of components, and utilizing the Pythagorean theorem and trigonometric functions, engineers can accurately calculate the resultant force and maintain the equilibrium and stability of engineering systems.

Equilibrium of Rigid Bodies

In the realm of statics, the equilibrium of rigid bodies is a fundamental concept that requires a deep understanding of the forces and moments acting on a body to ensure it remains in a state of rest or uniform motion. This concept is pivotal for engineers to design structures and mechanical systems that can withstand various loads without undergoing unintended motion. The conditions for equilibrium are derived from Newton's first law of motion, which states that a body will maintain its state of rest or uniform motion unless acted upon by an external force. For a rigid body to be in equilibrium, two main conditions must be satisfied: the sum of all horizontal forces must be zero, the sum of all vertical forces must be zero, and the sum of all moments about any point must also be zero. These conditions can be mathematically represented as $\sum F_x = 0$, $\sum F_y = 0$, and $\sum M_O = 0$, where F_x and F_y are the components of the forces in the x and y directions, respectively, and M_O represents the moments about point O.

To apply these conditions effectively, engineers must first identify all the forces acting on the body, including gravitational, applied, frictional, and reaction forces. Once these forces are identified, they are resolved into their horizontal and vertical components. This step is crucial for simplifying the analysis by allowing the use of scalar algebra to sum forces in each direction. For instance, if a beam is supported at two points and subjected to a uniform distributed load, the reaction forces at the supports must be calculated to ensure the beam remains in equilibrium. By applying the conditions for equilibrium, the engineer can set up equations based on the sum of forces in the vertical direction and the sum of moments about a point (often one of the supports) to solve for the unknown reactions.

The moment of a force about a point provides insight into the rotational effect of the force on the body and is calculated as the product of the force magnitude and the perpendicular distance from the point to the line of action of the force. This calculation is represented by $M = F \cdot d$, where M is the moment, F is the force, and d is the perpendicular distance. In analyzing the equilibrium of rigid bodies, considering moments is essential to ensure that the body does not rotate under the action of the applied forces.

In more complex scenarios, where forces are not perpendicular to their moment arms, the moment can be found using the cross product of the position vector r from the point to the point of application of the force and the force vector F, given by $M = r \times F$. This approach is particularly useful in three-dimensional equilibrium problems, where forces may act in any direction.

The concept of equilibrium extends beyond simply ensuring that a structure does not collapse. It is also used in the design of mechanical systems to ensure stability and proper function. For example, in the design of a crane, the equilibrium analysis ensures that the crane can lift loads without tipping over. This involves calculating the moments created by the weight of the load and the counterweight and ensuring they balance to maintain the crane in a stable equilibrium.

Frames and Trusses Analysis

Analyzing frames and trusses involves understanding the distribution of internal forces and ensuring structural stability, which is crucial for mechanical engineers preparing for the FE exam. Frames and trusses are fundamental components in many engineering structures, from bridges to buildings, and their analysis forms a core part of the statics section of the mechanical engineering discipline.

Frames are structures that consist of multiple members connected together, often forming a skeleton that supports loads. Unlike trusses, frames can support not only axial forces but also bending moments and shear forces. The analysis of a frame involves determining the reaction forces at supports and the internal forces within each member. This process typically begins with a free-body diagram of the entire structure to identify support reactions, followed by disassembling the frame into its individual members to analyze the forces acting on each. The equilibrium equations, $\sum F_x = 0$, $\sum F_y = 0$, and $\sum M = 0$, are applied at each joint or cut section to solve for unknown forces and moments.

Trusses are structures composed of slender members joined together at their ends to form a series of triangles. Trusses are primarily used to support loads and to transfer them efficiently through the structure to the supports. The analysis of trusses often employs two main methods: the method of joints and the method of sections. The **method of joints** starts with creating a free-body diagram for the truss and applying the equilibrium equations to each joint, assuming all members are in tension or compression and no bending moments are present. This method is particularly effective for determining the forces in all the members of the truss. On the other hand, the **method of sections** involves cutting through the truss to isolate a section and then applying the equilibrium equations to solve for the forces in the cut members. This method is useful for finding forces in specific members without having to analyze the entire truss.

Both methods rely on the fundamental principles of static equilibrium, where the sum of forces and moments in any direction equals zero. It's important to note that the assumption of pin-jointed members in trusses simplifies the analysis by eliminating moments at the joints, thus the members only carry axial forces.

In the context of the FE Mechanical exam, understanding the principles of statics applied to frames and trusses is essential. The ability to analyze these structures for internal forces and to ensure their stability under various loading conditions is a critical skill. This involves not only applying the equilibrium equations effectively but also understanding the physical behavior of the structure, such as determinacy and stability criteria. A statically determinate structure can be analyzed using only the equations of equilibrium, while a statically indeterminate structure requires additional compatibility equations or methods for analysis.

When preparing for the exam, focus on mastering the basic principles of equilibrium and the methods for analyzing both frames and trusses. Practice with a variety of problems to gain confidence in identifying the most efficient method of analysis for any given structure. Remember, the key to success in this area, as with all topics on the FE exam, is a solid understanding of the fundamentals and the ability to apply them to solve complex engineering problems.

Centroids and Moments of Inertia

The calculation of centroids and moments of inertia is a fundamental aspect of engineering mechanics, particularly within the realm of statics, which plays a crucial role in the analysis and design of structures and mechanical components. The centroid of a body represents the geometric center, a point where the body's shape, if made of a homogenous material, could be perfectly balanced on the tip of a pin. The moment of inertia, on the other hand, is a measure of an object's resistance to changes to its rotation. Both concepts are not only pivotal in understanding the distribution of mass and area but also in predicting the behavior of structures under various loads and moments.

Centroids are determined through the use of integral calculus, applying the principle that the centroid's coordinates, denoted as $(\bar{x}, \bar{y})$ for planar bodies, are the weighted average positions of the body's mass or area. For a simple geometric shape, the centroid can often be found at a point of symmetry. However, for more complex shapes or composite bodies, the centroid is found by dividing the shape into known geometries, calculating each section's centroid, and then finding the weighted average of these points. The formulas for determining the centroid of an area A are given by $\bar{x} = \frac{1}{A} \int x \, dA$ and $\bar{y} = \frac{1}{A} \int y \, dA$, where x and y are the coordinates of the infinitesimal area element dA.

Moments of inertia I, which are critical in predicting an object's angular motion and stability, are calculated by integrating over the body's mass or area distribution relative to a specific axis. For planar sections, moments of inertia are most commonly calculated about the centroidal axes, denoted as I_x and I_y. The general formulas for these moments are $I_x = \int y^2 \, dA$ and $I_y = \int x^2 \, dA$, where x and y are the distances of the element dA from the axes. Additionally, the polar moment of inertia, which is used in torsional analyses, is given by $J = I_x + I_y = \int r^2 \, dA$, where r is the radial distance from the axis of rotation to the element dA.

In engineering applications, the calculation of centroids and moments of inertia is not limited to simple geometric shapes but extends to complex structures through the use of composite areas or numerical integration techniques for irregular shapes. Understanding these concepts allows engineers to design structures and mechanical systems that are both efficient and resistant to bending, torsion, and buckling, ensuring safety and functionality.

The application of these principles is evident in the design of beams, where the moment of inertia influences the beam's bending stress and deflection. Similarly, in the analysis of columns, the centroidal axis is critical in determining buckling behavior. Thus, mastering the calculation of centroids and moments of inertia is essential for mechanical engineers preparing for the FE exam, as it underpins many of the statics problems they will encounter, from determining stress distributions in complex shapes to analyzing the stability of structures under various loading conditions.

Static Friction Concepts

Static friction is a force that resists the initiation of sliding motion between two surfaces in contact. It is a crucial concept in the realm of statics, especially for engineers preparing for the FE Mechanical exam. The magnitude of static friction depends on the nature of the surfaces in contact and the normal force pressing them together. The maximum static frictional force that can be applied before motion begins is given by $f_{max} = \mu_s N$, where μ_s is the coefficient of static friction and N is the normal force. This relationship highlights the proportional dependency of the frictional force on the normal force, a fundamental principle in understanding how objects will behave when subjected to various forces.

The coefficient of static friction is a dimensionless quantity that represents the frictional characteristics of the contact surfaces. It varies between different material pairs and is determined experimentally. For engineers, knowing the coefficient of static friction is essential for designing systems that require precise control over motion initiation, such as brake systems, clutches, and conveyor belts.

When analyzing problems involving static friction, it is important to recognize that the actual frictional force can vary from zero up to the maximum value f_{max}. This range allows for the static equilibrium of objects under varying loads without motion. The static frictional force adjusts within this range to exactly balance the applied force, up to its maximum limit, beyond which motion ensues. This self-adjusting nature of static friction is pivotal in solving statics problems, where the goal is often to determine the conditions under which an object will start to move or to calculate the forces at play when an object is on the verge of motion.

In practical applications, engineers use the concept of static friction to ensure stability and safety in designs. For example, the design of slopes or ramps involves calculating the maximum angle at which an object can rest without sliding, which directly depends on the coefficient of static friction between the object and the surface. Similarly, in the design of mechanical fasteners or components that rely on friction to transmit forces, understanding and applying the principles of static friction is key to ensuring that the components will perform as intended under load.

The analysis of static friction also extends to complex systems where multiple forces and moments are involved. In such cases, the equilibrium equations $\sum F_x = 0$, $\sum F_y = 0$, and $\sum M = 0$ are used in conjunction with the static friction formula to solve for unknowns. This integrated approach allows engineers to model real-world scenarios accurately, ensuring that their designs can withstand the forces they will encounter without unintended motion.

Understanding static friction is not only about preventing motion. It also plays a significant role in the controlled initiation of movement, such as in the launching of a spacecraft or the starting of a car, where overcoming static friction translates into the successful application of engineering principles to achieve desired outcomes. Therefore, mastering the concepts of static friction is indispensable for mechanical engineers aiming to excel in their careers and pass the FE Mechanical exam, as it underpins many of the statics problems they will encounter and provides a foundation for analyzing and designing a wide range of mechanical systems.

Chapter 7: Dynamics, Kinematics, Vibrations

Kinematics of Particles

Kinematics of particles is a fundamental aspect of dynamics that focuses on the motion of objects without considering the forces that cause such motion. It involves the study of the positions, velocities, and accelerations of particles as they move through space and time. The mathematical description of a particle's motion can be categorized into two main types: rectilinear and curvilinear motion. **Rectilinear motion** occurs when a particle moves along a straight path, characterized by a single spatial dimension, while **curvilinear motion** involves movement along a curved path, requiring the consideration of multiple spatial dimensions.

For rectilinear motion, the position of a particle can be described by its coordinate x along a straight line, with the positive direction usually defined by the problem's context. The **velocity** of the particle is the rate of change of its position with respect to time, given by the derivative of the position x with respect to time t, denoted as $v = \frac{dx}{dt}$. **Acceleration**, on the other hand, is the rate of change of velocity with respect to time, calculated as $a = \frac{dv}{dt} = \frac{d^2x}{dt^2}$. These relationships form the basis for analyzing rectilinear motion, allowing for the determination of a particle's position at any given time if its velocity and acceleration as functions of time are known.

Curvilinear motion requires a more complex analysis due to the particle's path's multidimensional nature. In this case, the motion is often described using vector notation, where the position of the particle is represented by a position vector **r** that originates from a reference point and terminates at the particle's location. The velocity **v** and acceleration **a** vectors are then defined as the first and second derivatives of the position vector with respect to time, respectively: $\mathbf{v} = \frac{d\mathbf{r}}{dt}$ and $\mathbf{a} = \frac{d\mathbf{v}}{dt} = \frac{d^2\mathbf{r}}{dt^2}$. For motion in two dimensions, these vectors can be broken down into their components along the x and y axes, providing a means to analyze the motion in terms of its horizontal and vertical components.

One of the key concepts in the kinematics of particles is the **path description**, which involves describing the motion of a particle along a predefined path, such as a circle or parabola. This often requires the use of parametric equations or specific path functions that relate the particle's position coordinates to each other and to time. For example, in circular motion, the position of a particle moving along the circumference of a circle can be described using trigonometric functions that relate the particle's x and y coordinates to the circle's radius and the angle swept by the radius vector over time.

Grasping the kinematics of particles is essential for addressing various engineering challenges, including the design of mechanical systems with moving components and the analysis of projectiles and satellite trajectories. Utilizing kinematic principles allows engineers to forecast the subsequent positions, velocities, and accelerations of moving objects, create control systems that manage their motion, and evaluate how different motion parameters influence the overall performance of mechanical and dynamic systems.

Kinetic Friction

Kinetic friction is a force that acts between moving surfaces and opposes the direction of motion. It is a critical concept in dynamics, particularly when analyzing the motion of objects in contact with each other. Unlike static friction, which prevents motion from initiating, kinetic friction operates when surfaces are already in motion relative to each other. The magnitude of kinetic friction can be calculated using the formula $f_k = \mu_k N$, where f_k is the force of kinetic friction, μ_k is the coefficient of kinetic friction, and N is the normal force exerted by the surface on the object. The coefficient of kinetic friction is a dimensionless value that depends on the materials of the contacting surfaces and is typically less than the coefficient of static friction for the same materials.

The normal force N is perpendicular to the contact surface and is equal in magnitude and opposite in direction to the force that the object applies on the surface. In many cases, this normal force is simply the weight of the object if the surface is horizontal and there are no other vertical forces acting on the object. However, in scenarios involving inclined planes or other vertical forces, the normal force must be calculated taking into account all the forces perpendicular to the surface.

The work done against kinetic friction when an object moves over a distance d can be expressed as $W = f_k d$, where W is the work done. This work results in the dissipation of energy, typically in the form of heat, which is an important consideration in the design of mechanical systems to prevent overheating and wear.

In practical applications, engineers must account for kinetic friction when designing mechanisms that involve sliding or rolling motion. For example, in the design of bearings, gears, and wheels, minimizing kinetic friction is crucial to reduce energy losses and extend the lifespan of these components. This is often achieved through material selection, lubrication, and surface treatments that reduce the coefficient of kinetic friction.

Moreover, kinetic friction plays a significant role in the analysis of vehicle dynamics, particularly in the context of braking systems. The ability of a vehicle to stop within a certain distance depends on the kinetic friction between the tires and the road surface. Understanding the factors that affect kinetic friction, such as road condition, tire material, and vehicle speed, is essential for ensuring the safety and performance of automotive vehicles.

In the realm of dynamics, kinematics, and vibrations, the study of kinetic friction is integral to predicting the motion of systems under the influence of external forces. It allows engineers to model the behavior of mechanical systems accurately and design solutions that optimize performance while mitigating the negative effects of friction. Whether analyzing the stability of structures subject to dynamic loads or designing efficient machines and vehicles, a thorough understanding of kinetic friction is indispensable for mechanical engineers preparing for the FE Mechanical exam.

Newton's Second Law for Particles

Newton's Second Law for Particles, a cornerstone of dynamics, articulates the relationship between the forces acting on a particle, its mass, and the resulting acceleration. This law is succinctly captured by the equation $F = ma$, where F represents the net force acting on the particle, m is the mass of the particle, and a is the acceleration produced by the force. The law posits that the acceleration of a particle is directly proportional to the net force acting on it and inversely proportional to its mass. This principle is pivotal in analyzing and predicting the motion of particles under the influence of various forces, making it indispensable for engineers preparing for the FE Mechanical exam.

The application of Newton's Second Law extends beyond simple linear motion to include rotational dynamics and systems of particles. In the context of particle dynamics, it's crucial to consider all forces acting on the particle, including gravitational, normal, frictional, and applied forces. The net force is the vector sum of these forces, and it dictates the direction and magnitude of the particle's acceleration. For instance, in the case of a particle sliding down an inclined plane, both gravitational force components along the incline and frictional forces opposing the motion must be accounted for to accurately apply Newton's Second Law.

When analyzing problems involving Newton's Second Law, it's essential to adopt a systematic approach: start by identifying and diagramming all forces acting on the particle, choose an appropriate coordinate system to simplify the analysis, and apply the law in vector form to solve for unknown quantities. This often involves breaking down the net force and acceleration into their component vectors, particularly in scenarios involving motion in multiple dimensions.

The law's utility is further demonstrated in its ability to model the dynamics of particles under a variety of conditions, including variable mass systems such as rockets expelling fuel. In such cases, the law is applied considering the changing mass of the system, leading to differential equations that describe the motion of the particle or system over time.

Moreover, Newton's Second Law forms the basis for the development of more complex concepts in dynamics, such as the principles of work and energy, impulse and momentum, and the analysis of vibrating systems. These concepts, while derived from the second law, provide alternative methods for solving dynamics problems that might be more suitable depending on the specific conditions of the problem at hand.

In preparing for the FE Mechanical exam, understanding and applying Newton's Second Law for Particles is fundamental. It not only enables the analysis of particle motion but also lays the groundwork for understanding more complex systems and phenomena in dynamics, kinematics, and vibrations. Mastery of this law, along with a solid grasp of the underlying physics and the ability to apply mathematical principles to solve problems, is essential for success on the exam and in professional practice.

Work-Energy of Particles

The work-energy principle is a fundamental concept in physics that relates the work done on a particle to its kinetic energy. This principle is particularly useful in analyzing the motion of particles where forces are applied, and it simplifies the process of solving mechanical problems. To understand the work-energy principle, it's essential to first define work and energy in the context of particle motion.

Work (W) is defined as the product of the force (F) applied to a particle and the displacement (d) of the particle in the direction of the force. Mathematically, it is expressed as $W = F \cdot d \cdot \cos(\theta)$, where θ is the angle between the force and the displacement vectors. It's important to note that work is a scalar quantity and its unit in the International System of Units (SI) is the Joule (J).

Kinetic energy (KE), on the other hand, is the energy that a particle possesses due to its motion. It is given by the equation $KE = \frac{1}{2}mv^2$, where m is the mass of the particle and v is its velocity. Like work, kinetic energy is measured in Joules (J).

The work-energy theorem states that the work done by all forces acting on a particle is equal to the change in its kinetic energy. This can be written as $W_{total} = \Delta KE = KE_{final} - KE_{initial}$. This theorem is powerful because it allows engineers to analyze particle motion without directly solving Newton's second law of motion, which requires consideration of the mass and acceleration of the particle.

In practical terms, the work-energy principle can be applied to various engineering problems, such as determining the speed of a roller coaster at the bottom of a hill, calculating the work required to lift an elevator to a certain height, or analyzing the impact forces in a car crash. By understanding the relationship between work and energy, engineers can design systems that are more efficient, safe, and effective.

One of the key advantages of using the work-energy principle is its ability to simplify complex problems where multiple forces, including friction and air resistance, act on a particle over a distance. Instead of analyzing each force separately, the principle allows for a holistic approach by considering the total work done and the resulting change in kinetic energy.

Impulse-Momentum of Particles

The impulse-momentum theorem is a fundamental principle in dynamics that relates the impulse applied to a particle to the change in its momentum. This theorem is particularly useful for analyzing situations where forces act on particles over a short time interval, such as collisions or explosions. The impulse J applied to a particle is the product of the force F acting on the particle and the time interval Δt during which the force acts. Mathematically, impulse is expressed as $J = F\Delta t$. Momentum, on the other hand, is the product of the mass m of the particle and its velocity v. The momentum of a particle is given by $p = mv$.

The impulse-momentum theorem states that the impulse applied to a particle is equal to the change in its momentum. This can be written as $J = \Delta p = p_{final} - p_{initial}$, where p_{final} is the momentum of the particle at the end of the time interval, and $p_{initial}$ is the momentum at the beginning of the time interval. This theorem is a powerful tool for solving engineering problems because it allows for the analysis of particle motion without the need to directly solve differential equations that describe the particle's motion.

In practical applications, the impulse-momentum theorem can be used to analyze impact forces in collisions. For example, in the case of a perfectly elastic collision, the kinetic energy and momentum are conserved. By applying the impulse-momentum theorem, engineers can determine the velocities of colliding particles after the collision if the initial velocities and the mass of the particles are known. Similarly, in inelastic collisions, where the particles stick together after the collision, the theorem can be used to find the final velocity of the combined mass.

Another important application of the impulse-momentum theorem is in the analysis of variable force impacts, such as a hammer striking a nail. The force exerted by the hammer on the nail varies over the duration of the impact. By integrating the force over the time of impact, the total impulse delivered to the nail can be calculated, which in turn allows for the determination of the nail's velocity immediately after the impact.

The theorem also finds application in sports engineering, where the performance of athletes can be analyzed and improved by studying the impulse applied during motions such as jumping, throwing, or striking. By maximizing the impulse delivered to an object, such as a ball, athletes can enhance their performance in sports activities.

For FE Mechanical exam candidates, understanding the impulse-momentum theorem is crucial for solving problems related to particle dynamics. Mastery of this concept enables candidates to efficiently tackle questions involving collisions, impacts, and other scenarios where forces act on particles over time. It is essential for candidates to be familiar with the mathematical formulation of the theorem and its applications in various engineering contexts, as this knowledge is directly applicable to questions they may encounter on the exam.

Kinematics of Rigid Bodies

Kinematics of rigid bodies is a fundamental aspect of mechanical engineering that deals with the motion of objects without considering the forces that cause this motion. This section delves into the analysis of both translational and rotational movements, which are pivotal for understanding the dynamics of mechanical systems. In the realm of translational motion, a rigid body is assumed to move in a path where all its parts travel the same distance in the same direction and in the same time interval. This type of motion can be described using the equations of linear motion, where the position x, velocity v, and acceleration a of the body are related to time t. The basic equations governing translational motion are $v = \dfrac{dx}{dt}$ and $a = \dfrac{dv}{dt}$, where dx represents a small change in position and dt represents a small change in time.

Rotational motion, on the other hand, involves a body rotating about an axis. Each point in the body follows a circular path, with all circles lying in planes perpendicular to the axis and having common centers along the axis. The angular position θ, angular velocity ω, and angular acceleration α are the rotational analogs of position, velocity, and acceleration in translational motion. The relationships between these quantities are given by $\omega = \frac{d\theta}{dt}$ and $\alpha = \frac{d\omega}{dt}$, where $d\theta$ is a small change in angular position. It's crucial to understand that while the equations of motion for translational and rotational movements are analogous, they apply to different physical quantities and units.

For a comprehensive analysis of rigid body motion, it's essential to combine translational and rotational kinematics. This is particularly relevant in cases where an object undergoes both types of motion simultaneously, such as a rolling wheel. In such scenarios, the point of contact between the wheel and the surface is momentarily at rest due to the frictional force, leading to a condition known as pure rolling motion. The velocity of any point on the wheel can be determined by vectorially adding the translational velocity of the center of mass and the rotational velocity around the center of mass. This approach allows for the determination of the velocity and acceleration of any point on a rotating body, providing a detailed understanding of its motion.

Furthermore, the concept of the moment of inertia plays a critical role in the analysis of rotational motion. The moment of inertia I of a body about a given axis describes how its mass is distributed relative to the axis and affects the body's resistance to changes in rotational motion. The moment of inertia is defined as $I = \int r^2 dm$, where r is the distance from the axis of rotation, and dm is an infinitesimal mass element of the body. This property is pivotal in calculating the kinetic energy of a rotating body, given by $K = \frac{1}{2}I\omega^2$, and in applying Newton's second law for rotation, $\tau = I\alpha$, where τ is the torque applied to the body.

The study of kinematics of rigid bodies includes both translational and rotational movements, forming a basis for analyzing the motion of mechanical systems. A grasp of linear and angular motion principles allows engineers to predict the behavior of mechanical systems under various conditions, design more efficient and effective machines, and address complex engineering problems related to motion.

Kinematics of Mechanisms

In the realm of mechanical engineering, the study of **kinematics of mechanisms** is pivotal for understanding how mechanical systems translate input forces and movements into desired outputs. This analysis is foundational for the design and optimization of various mechanical components and systems, including gears, levers, and linkages. The primary focus here is on the motion itself, rather than the forces that produce the motion, which is the domain of dynamics.

Mechanisms are composed of rigid bodies, known as links, connected by joints to form a closed chain or structure. The movement of these links relative to each other is of particular interest. The simplest form of a mechanism, the four-bar linkage, provides a fundamental example. In a four-bar linkage, four rigid links are connected in a loop by four joints. The analysis of this system involves determining the positions, velocities, and accelerations of the links as the mechanism moves through its range of motion.

To analyze the motion of mechanisms, one must first identify the **input link**, often driven by a motor or another power source, and the **output link**, which performs the desired task. The motion of the input link is known, and the goal is to determine the motion of the output link. This process involves the application of **geometric**, **kinematic equations**, and sometimes **graphical methods** to solve for the unknown positions, velocities, and accelerations of the mechanism's components.

Geometric methods involve using the geometric constraints of the mechanism to solve for the positions of the links. This can be done using trigonometric relationships and the law of cosines for triangles formed by the links of the mechanism. For example, in a four-bar linkage, if the lengths of all the links and the position of one link (the input) are known, the positions of the other links can be determined using geometric constructions.

Velocity analysis in mechanisms often employs the **relative velocity method**, where the velocity of a point on a link relative to another point on the same or another link is considered. This method utilizes the vector nature of velocity, where the velocity of a point on a link is the sum of the translational velocity of the link and the rotational velocity around the pivot point. The **instantaneous center of rotation** is a powerful concept in velocity analysis, providing a straightforward way to determine velocities without resorting to complex calculations.

Acceleration analysis further complicates the picture, as it involves not only the rates of change of the velocities but also the directions of these changes. The **tangential and normal components** of acceleration must be considered, especially in links undergoing circular motion.

The acceleration of a point on a link can be determined by analyzing the link's angular acceleration and the tangential and radial accelerations resulting from this motion.

In more complex mechanisms, such as those found in robotic arms or automotive engines, the analysis may require advanced mathematical models and computational tools. **Computer-aided design (CAD) software** and **multibody dynamics simulation** tools are often employed to model and analyze the motion of mechanisms accurately. These tools allow for the simulation of the mechanism's motion, providing insights into the velocities and accelerations of various components throughout the mechanism's operation.

For FE Mechanical exam candidates, mastering the principles of kinematics of mechanisms is essential for solving problems related to the design and analysis of mechanical systems, making it a critical area of study in preparation for the exam.

Newton's Second Law for Rigid Bodies

Newton's Second Law for Rigid Bodies is a cornerstone of dynamics, providing a framework for understanding how forces affect the motion of rigid bodies. This law states that the force acting on a rigid body is equal to the mass of the body multiplied by its acceleration. However, when dealing with rigid bodies, it's crucial to consider not only the linear but also the angular aspects of motion. The law can be expressed in two parts: one for linear motion and the other for rotational motion. For linear motion, the law is represented as $F = ma$, where F is the net force acting on the body, m is the mass of the body, and a is the linear acceleration. This equation implies that the acceleration of a body is directly proportional to the net force acting on it and inversely proportional to its mass.

For rotational motion, Newton's Second Law is adapted to account for the body's tendency to rotate about an axis, which is expressed as $\tau = I\alpha$, where τ represents the net torque acting on the body, I is the moment of inertia, and α is the angular acceleration. The moment of inertia is a measure of an object's resistance to changes in its rotational motion, analogous to mass in linear motion. It depends not only on the mass of the object but also on how that mass is distributed relative to the axis of rotation. The further the mass is from the axis, the higher the moment of inertia and the more torque is required to achieve the same angular acceleration.

Applying Newton's Second Law to rigid bodies involves analyzing the forces and torques acting on the body and determining the resulting linear and angular accelerations. This analysis is fundamental in mechanical engineering, as it allows engineers to predict the behavior of mechanical systems under various loads and forces. For instance, in designing a crane, engineers must ensure that the torques generated by the load do not exceed the crane's capacity to maintain stability and control. Similarly, in automotive engineering, understanding how forces and torques affect vehicle dynamics is crucial for designing safer and more efficient vehicles.

In practical applications, engineers often use free-body diagrams to visualize the forces and moments acting on a rigid body. These diagrams help in breaking down complex systems into simpler components, making it easier to apply Newton's Second Law. By summing up all the forces and setting them equal to ma and summing all the torques and setting them equal to $I\alpha$, engineers can solve for unknowns such as accelerations, forces, or torques. This process is essential in the design and analysis of mechanical systems, from simple machines to complex structures.

Moreover, the application of Newton's Second Law for rigid bodies extends beyond static systems to include dynamic situations where forces and torques vary over time. In such cases, differential equations are often used to model the system's behavior, allowing engineers to predict how rigid bodies will respond to varying conditions over time. This dynamic analysis is critical in many areas of engineering, including robotics, aerospace, and biomechanics, where understanding the effects of forces and torques on motion can lead to innovations and improvements in design and functionality.

Work-Energy Methods for Rigid Bodies

The work-energy principle for rigid bodies extends the concept of work and energy from particles to objects that do not deform under the influence of forces. This principle is particularly useful in mechanical engineering for analyzing the motion of objects where both translational and rotational movements are involved. The total mechanical energy of a rigid body is the sum of its kinetic energy due to translation, kinetic energy due to rotation, and potential energy. The work-energy theorem for rigid bodies states that the work done by external forces on a rigid body is equal to the change in its total mechanical energy.

Kinetic Energy due to Translation is given by $KE_{trans} = \frac{1}{2}mv^2$, where m is the mass of the rigid body and v is the velocity of its center of mass. This form is identical to the kinetic energy of a particle, reflecting the fact that translational motion of a rigid body can be considered as the motion of its center of mass.

Kinetic Energy due to Rotation about a fixed axis is expressed as $KE_{rot} = \frac{1}{2}I\omega^2$, where I is the moment of inertia of the body about the axis of rotation, and ω is the angular velocity. The moment of inertia is a measure of an object's resistance to changes in its rotational motion, analogous to mass in linear motion.

Potential Energy of a rigid body is typically due to gravitational forces and is calculated as $PE = mgh$, where m is the mass of the body, g is the acceleration due to gravity, and h is the height of the body's center of mass above a reference level.

The **Work-Energy Theorem** for a rigid body can be written as $W_{ext} = \Delta KE_{trans} + \Delta KE_{rot} + \Delta PE$, where W_{ext} is the work done by external forces on the body. This theorem is a powerful tool for solving engineering problems because it allows for the analysis of the body's motion without directly solving the equations of motion derived from Newton's second law.

In practical applications, engineers use the work-energy principle to design and analyze systems such as rotating machinery, vehicles, and structures subjected to dynamic loads. For example, in the analysis of a wind turbine, the work-energy principle can be applied to determine the kinetic energy imparted to the rotor by the wind, which is then converted into electrical energy. Similarly, in the crash analysis of vehicles, the principle is used to calculate the work done by impact forces and the resulting changes in kinetic and potential energies of the vehicle components.

The work-energy principle also simplifies the analysis of systems with non-conservative forces, such as friction. In these cases, the work done by non-conservative forces is accounted for separately, and the principle is modified to include the work done by these forces, leading to $W_{ext} + W_{nc} = \Delta KE_{trans} + \Delta KE_{rot} + \Delta PE$, where W_{nc} represents the work done by non-conservative forces.

In conclusion, the work-energy methods for rigid bodies provide a comprehensive framework for analyzing the motion of objects in mechanical engineering. By understanding and applying these principles, engineers can efficiently solve complex problems related to the dynamics of rigid bodies, enhancing their ability to design and optimize mechanical systems for a wide range of applications.

Impulse-Momentum in Rigid Body Dynamics

The impulse-momentum principle for rigid bodies is a fundamental concept in dynamics that relates the impulse applied to a rigid body to the change in its momentum. This principle is particularly useful in analyzing situations where forces are applied over short time intervals, such as collisions or explosions. The impulse J experienced by a rigid body is the integral of the net force F applied to it over the time interval Δt, expressed as $J = \int_{t_1}^{t_2} F\, dt$. The momentum of a rigid body, on the other hand, is the product of its mass m and its velocity v, given by $p = mv$. For rotational motion, the angular momentum L about a point is the product of the moment of inertia I and the angular velocity ω, expressed as $L = I\omega$.

When considering the impulse-momentum principle for rigid bodies, it's essential to differentiate between linear and angular components. For linear motion, the change in momentum Δp of the body is directly proportional to the impulse applied, which can be represented as $\Delta p = J$. This implies that the final momentum p_f of the body is equal to its initial momentum p_i plus the impulse, or $p_f = p_i + J$. In the context of angular motion, the change in angular momentum ΔL is proportional to the angular impulse applied, leading to a similar relationship: $\Delta L = J_\theta$, where J_θ is the angular impulse. This results in the final angular momentum L_f being equal to the initial angular momentum L_i plus the angular impulse, or $L_f = L_i + J_\theta$.

The application of the impulse-momentum principle in rigid body dynamics allows for the analysis of various engineering problems, such as determining the effects of forces applied during a short time interval on the motion of a rigid body. This principle is crucial in the design and analysis of mechanical systems subjected to dynamic loads, including vehicles in crash scenarios, machinery components during sudden starts or stops, and sports equipment upon impact.

In analyzing real-world problems using the impulse-momentum principle, engineers often make use of the conservation of momentum in the absence of external forces. This is particularly relevant in collision analysis, where the total momentum before the collision is equal to the total momentum after the collision, assuming no external impulse is applied. By understanding the impulse-momentum relationship, engineers can predict the post-collision velocities of colliding bodies, design safer automotive structures, and optimize the performance of mechanical systems.

Furthermore, the impulse-momentum principle is instrumental in the study of vibrations and impacts in mechanical systems, providing insights into the behavior of systems under transient forces. By applying this principle, engineers can design systems that minimize undesirable vibrations and absorb impacts efficiently, enhancing the durability and performance of mechanical components.

Free and Forced Vibrations Analysis

In the realm of mechanical engineering, understanding the dynamics of **free and forced vibrations** is crucial for designing systems that can withstand or exploit these phenomena. Free vibrations occur when a system oscillates in the absence of external forces after being initially disturbed. The behavior of such a system is governed by its natural frequency, denoted by ω_n, which is a function of the system's mass m and stiffness k, given by the formula $\omega_n = \sqrt{\frac{k}{m}}$. This fundamental relationship highlights the intrinsic properties that dictate how a system will respond to an initial displacement or force.

Forced vibrations, on the other hand, result when an external time-varying force is applied to a system, compelling it to oscillate at the frequency of the force rather than at its natural frequency. The analysis of forced vibrations is essential for understanding how systems respond to periodic driving forces, which can vary in magnitude and frequency. The response of a system to forced vibrations is characterized by its amplitude and phase angle, both of which depend on the driving frequency relative to the system's natural frequency.

A critical aspect of forced vibration analysis is the concept of **resonance**, which occurs when the frequency of the external force matches the natural frequency of the system. At resonance, even small periodic forces can induce large amplitude oscillations due to the constructive interference of the energy input into the system. The mathematical representation of this phenomenon is captured by the equation for the amplitude of the response, A, as a function of the driving frequency, ω, given by $A(\omega) = \dfrac{F_0}{\sqrt{(\omega_n^2 - \omega^2)^2 + (2\zeta\omega_n\omega)^2}}$, where F_0 is the magnitude of the external force, and ζ is the damping ratio of the system. This equation underscores the significant impact of damping on the system's response, particularly near resonance, where damping can mitigate potentially destructive oscillations.

The analysis of vibrations, both free and forced, extends beyond theoretical constructs to practical applications in engineering design and analysis. For instance, engineers must consider the effects of resonance in the design of bridges, buildings, and mechanical components to prevent structural failures. Similarly, the principles of vibration analysis are applied in the design of automotive suspension systems to enhance ride comfort and vehicle stability by minimizing the impact of road irregularities.

Moreover, the study of vibrations is not limited to macroscopic systems. In the microscale, engineers leverage vibration analysis in the design of MEMS (MicroElectroMechanical Systems) devices, where resonance phenomena can be exploited for sensing and actuation purposes. The precise control of vibrational modes in these devices is critical for their functionality and efficiency.

The analysis of free and forced vibrations is essential for comprehending the behavior of mechanical systems under different conditions. Mastery of these principles enables engineers to create systems that are more resilient and efficient, anticipate and address potential vibration-related issues, and utilize resonance effectively in various applications, from structural engineering to microscale devices. This expertise is crucial not only for success on the FE Mechanical exam but also for progress in the mechanical engineering field, where dynamics, kinematics, and vibrations are applied to tackle complex real-world challenges.

Chapter 8: Mechanics of Materials

Shear and Moment Diagrams

Constructing shear and moment diagrams is a fundamental skill in the analysis of beam loading conditions, essential for understanding the internal forces and moments that act along the length of a beam. These diagrams provide a visual representation of how shear forces and bending moments vary along the beam, which is crucial for the design and analysis of structural elements. The process of constructing these diagrams involves several steps, starting with the identification of beam supports, applied loads, and reactions.

Shear Force Diagram (SFD): The shear force diagram illustrates how the shear force varies along the length of the beam. To construct an SFD, begin by calculating the reactions at the supports using equilibrium equations. Once the reactions are determined, plot the shear force at the left end of the beam, which is typically at one of the supports. Moving from left to right, the value of the shear force changes as you encounter applied loads or moments. For vertical loads, the shear force will increase or decrease abruptly, indicating a change in the internal shear force within the beam. The sign of the shear force indicates the direction, with downward forces considered negative and upward forces positive. The SFD is drawn step by step, segment by segment, reflecting these changes until reaching the opposite end of the beam.

Bending Moment Diagram (BMD): The bending moment diagram shows the variation of bending moment along the beam and is constructed following the completion of the SFD. Starting from the left, the bending moment at the supports is often zero, except in the case of fixed supports where moments can be applied. As you move along the beam, the bending moment changes linearly in regions under uniform load and remains constant under no load. The magnitude of the bending moment at any point is equal to the area under the shear force diagram up to that point. Points of zero shear force are of particular interest as they indicate potential locations of maximum bending moments. The BMD is plotted by calculating these moments at various points along the beam, taking into account the sign convention where moments causing sagging are considered positive and those causing hogging are negative.

The relationship between the load on the beam, the shear force, and the bending moment is integral to understanding beam behavior. The derivative of the moment diagram with respect to the beam length gives the shear force, and similarly, the derivative of the shear force diagram gives the distributed load on the beam. This interrelation provides a methodical approach to analyzing beam problems, where the accuracy of the SFD directly influences the BMD.

In practice, the construction of shear and moment diagrams is not only a theoretical exercise but also a practical tool in engineering design. These diagrams allow engineers to identify critical points along the beam where the shear force and bending moments reach their maximum values, dictating the design requirements for beam dimensions, material selection, and reinforcement needs. Understanding the distribution of internal forces and moments is crucial for ensuring the structural integrity and safety of engineering projects, from simple beams to complex frameworks in buildings and bridges.

The process of constructing shear and moment diagrams requires a systematic approach, starting with a clear understanding of the loading conditions and support configurations. By breaking down the beam into segments and analyzing each segment for its contribution to the internal shear forces and bending moments, engineers can effectively design and analyze structural elements to meet safety and performance criteria. This analytical skill is not only essential for passing the FE Mechanical exam but also forms the foundation of structural analysis and design in mechanical engineering.

Stress Transformations and Mohr's Circle

Stress transformations are a fundamental concept in the mechanics of materials, essential for understanding how stress components change when the orientation of the coordinate system is altered. This knowledge is crucial for analyzing and designing components subjected to complex stress states. The transformation of plane stress and the use of Mohr's Circle for stress analysis are powerful tools in this regard. When a material element is subjected to a general state of plane stress, it experiences normal stresses σ_x and σ_y along the x and y axes, respectively, and a shear stress τ_{xy} acting on the xy plane. The orientation of the element can significantly influence the magnitude of these stress components.

To analyze the stress state at any arbitrary angle θ in the plane, stress transformation equations are employed. These equations are derived from the equilibrium of forces and moments on the element and are given by:

$$\sigma_{x'} = \frac{\sigma_x + \sigma_y}{2} + \frac{\sigma_x - \sigma_y}{2}\cos(2\theta) + \tau_{xy}\sin(2\theta)$$

$$\sigma_{y'} = \frac{\sigma_x + \sigma_y}{2} - \frac{\sigma_x - \sigma_y}{2}\cos(2\theta) - \tau_{xy}\sin(2\theta)$$

$$\tau_{x'y'} = -\frac{\sigma_x - \sigma_y}{2}\sin(2\theta) + \tau_{xy}\cos(2\theta)$$

These equations allow engineers to calculate the normal and shear stresses on planes at any orientation θ from the original coordinate axes. Understanding how to apply these equations is critical for identifying the maximum normal and shear stresses that a material element might experience, which in turn is essential for ensuring that designs are safe, efficient, and effective.

Mohr's Circle is a graphical method for performing stress transformations and finding principal stresses, maximum shear stresses, and the orientations of the planes on which these occur. To construct Mohr's Circle, the normal stress σ is plotted on the horizontal axis, and the shear stress τ is plotted on the vertical axis. The circle is then drawn with its center at $\frac{\sigma_x + \sigma_y}{2}$ on the σ axis and a radius equal to the square root of $\left(\frac{\sigma_x - \sigma_y}{2}\right)^2 + \tau_{xy}^2$. The points where the circle intersects the σ axis represent the principal stresses, σ_1 and σ_2, and the maximum shear stress is represented by the radius of the circle. The angle to the principal planes from the x-axis is found by measuring the angle from the x-axis to the line connecting the center of the circle to the point of interest and doubling it.

Mohr's Circle not only simplifies the process of finding principal and maximum shear stresses but also provides intuitive insights into the nature of stress states within materials. By analyzing the circle, engineers can quickly determine the critical stress conditions and the orientations at which these occur, which is invaluable for the design and analysis of mechanical components and structures.

Stress and Strain from Axial Loads

When a structural element is subjected to axial loads, the resulting stress and strain within the material are critical factors that must be accurately determined to ensure the integrity and safety of the structure. Axial loading refers to forces applied along the longitudinal axis of an element, causing it to either stretch (tension) or shorten (compression). The stress (σ) induced by axial loads is calculated by dividing the force (F) by the cross-sectional area (A) of the element, as given by the formula $\sigma = \frac{F}{A}$. This relationship highlights the direct proportionality between the applied force and the resulting stress, emphasizing the importance of understanding the material's capacity to withstand such stresses without failure.

Strain (ϵ), on the other hand, measures the deformation experienced by the material as a result of the applied stress and is defined as the change in length (ΔL) divided by the original length (L_0)

of the material, expressed as $\epsilon = \frac{\Delta L}{L_0}$. The strain provides insight into the extent of deformation, which is crucial for assessing whether the material will return to its original shape (elastic deformation) or undergo permanent deformation (plastic deformation).

The relationship between stress and strain for a material under axial loading is characterized by the material's modulus of elasticity (E), also known as Young's modulus. The modulus of elasticity is a measure of the material's stiffness and is defined as the ratio of stress to strain under elastic deformation conditions, $E = \frac{\sigma}{\epsilon}$. This parameter is fundamental in determining the material's response to applied loads and is essential for the design and analysis of structural elements to prevent failure under operational loads.

In engineering practice, the analysis of stress and strain caused by axial loads involves considering both the material properties and the geometric characteristics of the structural element. For example, elements with larger cross-sectional areas will experience lower stress under the same axial load compared to elements with smaller cross-sectional areas. Similarly, materials with higher modulus of elasticity values are stiffer and less prone to deformation under axial loads, which is a desirable property for structural applications where minimal deformation is required.

Furthermore, the concept of safety factors is employed to ensure that the design stress does not exceed the material's yield strength, the stress at which a material begins to deform plastically. By applying a safety factor, engineers can design structures that have adequate strength to withstand unexpected loads or uncertainties in the material properties without catastrophic failure.

The analysis of stress and strain resulting from axial loads is a crucial component of the mechanics of materials, forming the foundation for the design and assessment of structural elements. The relationship between applied loads, material properties, and the resulting stress and strain is vital for ensuring the safety, reliability, and performance of engineering structures. Careful consideration of these factors enables engineers to design structures that fulfill the necessary specifications and endure operational demands, thereby advancing the field of mechanical engineering and contributing to the creation of safe and efficient engineering solutions.

Stress and Strain from Bending Loads

When analyzing the effects of bending loads on beams and structures, it is essential to understand the concepts of stress and strain that these loads induce. Bending stress, or flexural stress, arises from the external load applied perpendicularly to the longitudinal axis of the beam, causing the beam to bend. The bending moment, a measure of the internal moment that resists the applied load, is crucial for calculating the bending stress. The linear distribution of stress across the height of the beam's cross-section is one of the fundamental principles of bending stress analysis. The maximum stress occurs at the outermost fibers of the beam, furthest from the neutral axis, where the neutral axis itself experiences no stress. This distribution is governed by the flexural formula:

$$\sigma = \frac{My}{I}$$

where σ is the bending stress, M is the bending moment at the cross-section under consideration, y is the distance from the neutral axis to the point where the stress is being calculated, and I is the moment of inertia of the beam's cross-section about the neutral axis. The moment of inertia is a geometric property that reflects the cross-section's resistance to bending and is dependent on the shape of the cross-section.

Strain in a beam under bending loads also varies linearly from the neutral axis, with the maximum strain occurring at the surface furthest from the neutral axis. The relationship between stress and strain in the elastic region of the material is defined by Hooke's Law, which states that stress is proportional to strain. The modulus of elasticity, or Young's modulus E, is the constant of proportionality and is a material property that measures the stiffness of the material:

$$\epsilon = \frac{\sigma}{E}$$

where ϵ is the strain. In the context of bending, the curvature of the beam, κ, is related to the bending moment M and the modulus of elasticity E through the equation:

$$\kappa = \frac{M}{EI}$$

This equation highlights the interplay between the material properties (E), the geometric properties (I), and the applied loads (M) in determining the deformation of the beam under bending loads.

The analysis of stress and strain due to bending loads is not complete without considering the beam's support conditions and loading configuration, as these factors significantly influence the bending moment distribution along the length of the beam. For instance, a simply supported beam with a concentrated load at its midpoint will have a different moment distribution compared to a cantilever beam with a uniformly distributed load. The calculation of bending moments and shear forces at various points along the beam is facilitated by constructing shear force and bending moment diagrams, which serve as foundational tools for understanding the internal actions within the beam.

Moreover, real-world applications often involve beams with complex cross-sections, materials with non-linear stress-strain relationships, or conditions leading to plastic deformation. In such cases, advanced analytical or numerical methods, such as finite element analysis (FEA), may be required to accurately predict the behavior of the structure under bending loads. These methods allow for a more detailed examination of the stress and strain distribution, taking into account factors such as material heterogeneity, non-linear material behavior, and geometric non-linearities due to large deformations.

In engineering practice, ensuring the structural integrity and safety of components subjected to bending loads involves not only the accurate calculation of stress and strain but also the application of design criteria, such as the factor of safety and allowable stress. These criteria account for uncertainties in material properties, loading conditions, and the potential for imperfections in the structure, thereby providing a margin of safety against failure.

Torsional Stress and Strain Effects

Torsional stress and strain are critical factors in the design and analysis of shafts and other cylindrical elements subjected to twisting moments. This type of stress arises when a torque is applied, causing the element to twist along its longitudinal axis. The analysis of torsional stress is essential for ensuring the structural integrity and functionality of mechanical components such as drive shafts, axles, and power transmission gears.

The fundamental equation for calculating torsional stress τ in a circular shaft is given by $\tau = \dfrac{Tc}{J}$, where T represents the applied torque, c is the outer radius of the shaft, and J is the polar moment of inertia of the cross-section. The polar moment of inertia for a solid circular shaft can be calculated using $J = \dfrac{\pi d^4}{32}$, where d is the diameter of the shaft. For hollow shafts, the equation adjusts to $J = \dfrac{\pi(d_o^4 - d_i^4)}{32}$, with d_o and d_i representing the outer and inner diameters, respectively.

Torsional strain, on the other hand, measures the deformation of the shaft due to the applied torque. It is a dimensionless quantity that describes the angle of twist per unit length of the shaft. The angle of twist, θ, in radians, can be found using $\theta = \dfrac{Tl}{GJ}$, where l is the length of the shaft and G is the shear modulus of the material. The shear modulus is a material property that quantifies its ability to deform under shear stress.

The relationship between torsional stress and strain is governed by the material's shear modulus, indicating that the torsional stiffness of a shaft is directly proportional to its material's shear modulus and its geometric properties. This relationship highlights the importance of material selection and geometric design in the engineering of components subjected to torsional loads.

In practical applications, engineers must also consider the effects of stress concentrations, which occur at points where there is a sudden change in the cross-section of the shaft, such as keyways, splines, or holes. Stress concentration factors can significantly increase the local stress above the average stress calculated using the equations mentioned above. Advanced topics such as these require careful consideration and are often addressed using finite element analysis (FEA) for more accurate predictions of stress and strain distributions.

Understanding the principles of torsional stress and strain is crucial for the design of mechanical systems that are safe, efficient, and reliable. Proper analysis ensures that components will withstand operational loads without failure, contributing to the overall durability and performance of mechanical systems.

Stress and Strain from Shear Forces

In the analysis of **beam sections** under shear forces, understanding the distribution and impact of **shear stress** τ and **shear strain** γ becomes paramount. Shear stress arises when external forces are applied parallel to the material's surface, causing layers of the material to slide past each other. The formula to calculate shear stress in a beam subjected to a vertical shear force V is given by $\tau = \dfrac{VQ}{It}$, where Q is the first moment of area above the point where the shear stress is being calculated, I is the moment of inertia of the entire cross-sectional area about the neutral axis, and t is the thickness of the material at the point of interest.

The **first moment of area**, Q, is a measure of the distribution of the area's shape relative to an axis, calculated as the integral of the area times the perpendicular distance from the axis $\int y dA$. This concept is crucial in determining how the shape of a beam or structural element influences the shear stress distribution across its cross-section. The **moment of inertia**, I, on the other hand, represents the beam's resistance to bending and is calculated based on the geometry of the cross-section.

Shear strain, represented by γ, describes the deformation of the material under shear stress. It is the angular distortion resulting from the applied force, defined as the change in angle between two originally perpendicular lines. In the context of shear, strain can be approximated as $\gamma = \frac{\tau}{G}$, where G is the shear modulus of the material, a property that quantifies the material's ability to deform under shear stress.

The relationship between shear stress and shear strain is linear for elastic materials, following **Hooke's Law** for shear $\tau = G\gamma$, indicating that the deformation is reversible and proportional to the applied force up to the yield point of the material. This linear behavior simplifies the analysis and design of structural elements by enabling engineers to predict the performance of materials under known loads.

In practical applications, especially in **beam sections**, the distribution of shear stress is not uniform across the depth of the beam. Near the neutral axis, where bending moments are highest, shear stress reaches its maximum value and diminishes towards the outer fibers of the beam. This non-uniform distribution is critical in the design and analysis of beams, as it affects the beam's strength and deflection under load.

Moreover, the concept of **shear flow** in built-up beams, such as those made from several materials or sections, plays a significant role in understanding how shear forces are transferred between components. Shear flow, q, defined as $q = \tau t$, where t is the width of the section, helps in designing fasteners and connectors by ensuring that shear forces are adequately shared among the components of the composite beam.

Thermal Stress and Strain

Thermal stress and strain arise when a material undergoes a temperature change, leading to expansion or contraction that is constrained. This phenomenon is critical in engineering applications where materials are subjected to varying thermal environments, potentially affecting their structural integrity and performance. The fundamental principle governing thermal stress, $\sigma_{thermal}$, can be expressed by the equation $\sigma_{thermal} = E\alpha\Delta T$, where E is the modulus of elasticity of the material, α is the coefficient of thermal expansion, and ΔT is the change in temperature. The coefficient of thermal expansion is a material property that quantifies how much a material expands per unit length for a unit change in temperature. It is crucial for engineers to understand and account for the effects of thermal expansion to prevent structural failures or to design components that can accommodate or utilize thermal expansion.

Thermal strain, $\epsilon_{thermal}$, associated with temperature changes, does not require the application of external forces and is calculated by $\epsilon_{thermal} = \alpha\Delta T$. This relationship highlights that thermal strain is directly proportional to the temperature change and the material's coefficient of thermal expansion. In scenarios where expansion is not constrained, this strain manifests as a change in dimensions of the material. However, in engineering applications, materials are often confined or connected to other materials with different expansion rates, leading to the development of thermal stresses.

In the context of composite materials or assemblies of different materials, the mismatch in coefficients of thermal expansion can lead to significant internal stresses when the temperature changes. Engineers must carefully select materials with compatible thermal expansion properties or design the structure to accommodate or relieve the stresses that may develop. For example, in bridges or railway tracks, expansion joints are used to allow for the expansion and contraction of materials, thereby preventing undue stress buildup.

The analysis of thermal stress is further complicated in anisotropic materials, where the coefficient of thermal expansion varies with direction within the material. In such cases, the thermal stress and strain must be evaluated in each principal direction, taking into account the material's anisotropic properties.

Moreover, in transient thermal conditions where temperature changes with time, the distribution of thermal stress and strain can vary throughout the structure. The analysis of such conditions requires solving the heat conduction equation to determine the temperature distribution and then calculating the resulting stress and strain distribution. Advanced computational methods, such as finite element analysis (FEA), are often employed to accurately model and analyze these complex scenarios.

Understanding the interplay between thermal stress and strain is essential for designing components and structures that operate reliably under varying temperatures. This includes everything from electronic devices, where thermal expansion can lead to failure of solder joints or other connections, to large-scale structures like pipelines or skyscrapers, where temperature variations can be significant. Engineers must account for thermal effects not only in the material selection and component design phase but also in the operational analysis to ensure longevity and reliability of the engineering systems.

Combined Loading Analysis

In the realm of mechanics of materials, understanding the behavior of structures or components under combined loading is crucial for accurate analysis and design. Combined loading refers to situations where a material or structural element is subjected to two or more types of loads simultaneously, such as axial, torsional, and bending loads. The complexity of combined loading scenarios necessitates a comprehensive approach to ensure the integrity and functionality of the component under various operational conditions.

When analyzing combined loading, the superposition principle is often employed. This principle allows the separate effects of each type of load to be analyzed independently before their effects are algebraically added to determine the overall response of the material or structure. However, it's imperative to note that superposition only applies when the material behavior remains linear and elastic.

For axial and bending loads, the stress in the material can be expressed as $\sigma = \frac{P}{A} + \frac{M_y}{I}y + \frac{M_z}{I}z$, where P represents the axial load, A is the cross-sectional area, M_y and M_z are the bending moments about the y and z axes respectively, I is the moment of inertia, and y and z are the distances from the neutral axis. This equation highlights how axial and bending stresses are superimposed to determine the total stress at any point within the material.

Torsional loading adds another layer of complexity, as it introduces shear stress perpendicular to the axial and bending stresses. The torsional stress, τ, for a circular shaft can be calculated using $\tau = \frac{Tc}{J}$, where T is the applied torque, c is the outer radius of the shaft, and J is the polar moment of inertia. In combined loading scenarios involving torsion, the resultant stress state may necessitate the use of more advanced theories, such as the von Mises yield criterion, to accurately predict yielding. The von Mises stress, σ_{vm}, is particularly useful as it combines the normal and shear stresses into a single equivalent stress value, given by $\sigma_{vm} = \sqrt{\sigma^2 + 3\tau^2}$, where σ is the normal stress from axial and bending loads, and τ is the shear stress from torsional load.

In practical applications, engineers must also consider the effects of stress concentrations, which can significantly amplify stresses in areas where there are abrupt changes in geometry, such as holes, notches, or fillets. The presence of stress concentrations in combined loading scenarios requires careful analysis, often involving numerical methods like finite element analysis (FEA), to accurately predict the stress distribution and identify potential failure points.

Moreover, real-world engineering components often experience complex loading conditions that vary over time, further complicating the analysis. For instance, a rotating shaft in machinery may be subjected to fluctuating torsional loads due to changes in operational speed, in addition to constant bending moments from attached weights and axial loads from tension or compression. In such cases, fatigue analysis becomes essential to assess the component's life expectancy under cyclic loading conditions.

The analysis of materials subjected to combined loading presents a complex challenge that necessitates a thorough grasp of material behavior, stress analysis, and failure theories. Engineers must skillfully apply the principles of superposition, take into account stress concentrations, and utilize advanced analytical or numerical methods to guarantee the safety, reliability, and performance of structures and components across various applications.

Deformation Calculation Methods

Deformations in materials under various loading conditions are quantified using several fundamental equations and concepts that are essential for the analysis and design of mechanical components. The calculation of deformations is crucial for ensuring that structures and components can withstand operational stresses without failure. This section delves into the methods used to calculate deformations, focusing on axial, bending, and torsional loads, which are prevalent in many engineering applications.

Axial Deformation is calculated using the formula $\delta = \frac{FL}{AE}$, where F is the axial force applied to the object, L is the original length of the object, A is the cross-sectional area, and E is the modulus of elasticity of the material. This equation is derived from Hooke's Law for linear elastic materials, which states that the strain (deformation per unit length) in a material is proportional to the stress applied to it. The modulus of elasticity, E, is a material property that measures its stiffness and is a critical factor in determining how much a material will deform under a given load.

Bending Deformation in beams is more complex due to the distribution of stresses across the beam's cross-section. The maximum deformation, or deflection, in a beam subjected to bending loads can be calculated using various methods, depending on the beam's support conditions and load distribution. For a simply supported beam with a concentrated load at the center, the maximum deflection, y_{max}, can be found using $\frac{FL^3}{48EI}$, where F is the load applied, L is the length of the beam, E is the modulus of elasticity, and I is the moment of inertia of the beam's cross-section about the neutral axis. The moment of inertia, I, is a geometric property that represents how the cross-section's area is distributed about the neutral axis, affecting the beam's resistance to bending.

Torsional Deformation of a cylindrical shaft subjected to a torque T is given by $\theta = \frac{TL}{GJ}$, where T is the applied torque, L is the length of the shaft, G is the shear modulus of the material, and J is the polar moment of inertia of the shaft's cross-section. The shear modulus, G, is a material property that indicates its ability to deform under shear stress. The polar moment of inertia, J, similar to the moment of inertia in bending, is a geometric property that affects the shaft's resistance to twisting. The angle of twist, θ, is a measure of how much one end of the shaft has rotated relative to the other end.

In addition to these fundamental loading conditions, engineers must often analyze **combined loading scenarios** where components are subjected to a mixture of axial, bending, and torsional loads. The superposition principle is employed in linear elastic regimes to calculate the total deformation as the sum of deformations due to each type of load, assuming that the material's response to each loading type is independent of the others. However, in real-world applications, the interaction between different types of stresses and deformations can be complex, requiring the use of numerical methods such as finite element analysis (FEA) for accurate predictions.

The calculation of deformations is a critical aspect of the design and analysis process in mechanical engineering. Understanding how materials and structures deform under various loads allows engineers to predict their behavior under operational conditions, ensuring that they can withstand the forces they will encounter without excessive deformation or failure. This knowledge is foundational for the design of safe, efficient, and durable mechanical systems and structures.

Column Buckling Behavior

Column buckling is a critical phenomenon in the mechanics of materials, particularly when dealing with the stability of structures subjected to compressive loads. It is essential to understand that columns buckle due to a loss of stability, which can occur even if the stress in the column has not reached the material's yield strength. This behavior is governed by the Euler Buckling Formula, which provides a mathematical model to predict the critical load at which a slender column will buckle under axial compression. The formula is expressed as

$$P_{cr} = \frac{\pi^2 EI}{(KL)^2}$$

, where P_{cr} is the critical load, E is the modulus of elasticity of the material, I is the moment of inertia of the cross-section about the axis of bending, L is the unsupported length of the column, and K is the column effective length factor, which depends on the conditions of end support. The effective length factor, K, varies with the boundary conditions; for example, a column with both ends pinned (hinged) has $K = 1$, while a column with one end fixed and the other free to move laterally has $K = 2$.

The critical stress associated with the critical load can be determined by dividing the critical load by the cross-sectional area A of the column, yielding

$$\sigma_{cr} = \frac{P_{cr}}{A} = \frac{\pi^2 E}{(KL/r)^2}$$

, where r is the radius of gyration of the column's cross-section, defined as $r = \sqrt{I/A}$. This relationship highlights the inverse square relationship between the critical stress and the slenderness ratio KL/r, indicating that as the slenderness ratio increases, the capacity of the column to carry axial loads without buckling decreases.

In practical engineering applications, the design of columns must consider the buckling criterion to ensure that columns are capable of supporting the expected loads without risk of sudden buckling failure. The selection of materials, cross-sectional shape, and column length must be optimized to increase the critical load and thus enhance the stability of the column under compressive forces. Additionally, real-world columns are rarely ideal and may have imperfections such as initial curvature, eccentric loading, or variations in cross-sectional area, which can reduce the critical load compared to the idealized Euler case. Therefore, design codes often include safety factors and empirical modifications to account for these imperfections and ensure the safe design of structural columns.

Moreover, the concept of buckling is not limited to columns but can also apply to plates and shells, where compressive stresses can lead to buckling in modes that differ from the simple column buckling. In these cases, the analysis becomes more complex, and advanced computational methods, such as finite element analysis (FEA), are employed to accurately predict buckling loads and modes.

Statically Indeterminate Systems

Statically indeterminate systems present a unique challenge in the field of mechanics of materials, as these systems have more unknowns than the available equilibrium equations can solve. This discrepancy arises because the number of reactions or internal forces exceeds the number of equilibrium equations, making the system indeterminate through equilibrium considerations alone. To address these systems, engineers must employ compatibility conditions, which are additional constraints that ensure the deformation of the structure meets certain criteria, allowing for a complete solution to be found.

Compatibility conditions are based on the physical behavior of materials and the geometry of the structure. They ensure that the deformation of the structure is continuous and consistent across the entire system. For example, in a beam supported by more than two supports, the deflection of the beam at the supports must be zero, and the slope of the deflection curve at points of continuity must be the same from both sides. These conditions are essential for solving statically indeterminate systems because they provide the additional equations needed to solve for the unknowns.

The general approach to solving statically indeterminate systems involves the following steps:
1. Determine the degree of static indeterminacy, which is the number of extra unknowns beyond what the equilibrium equations can solve.
2. Apply the equilibrium equations to the system to solve for as many unknowns as possible.
3. Develop compatibility conditions based on the geometry of the system and the deformation characteristics of the materials involved.
4. Use the compatibility conditions to set up additional equations that, together with the equilibrium equations, allow for the solution of all unknowns.

An essential tool in this process is the method of superposition, which allows the engineer to consider the effect of individual loads separately and then superimpose the results to find the total effect. This method is particularly useful in statically indeterminate systems because it simplifies the analysis of complex loading conditions.

For example, consider a beam subjected to a uniform load and supported by three supports. The reactions at the supports cannot be determined by equilibrium equations alone because there are three unknown reactions but only two independent equilibrium equations. By applying a compatibility condition, such as the deflection at the middle support must be zero, an additional equation is obtained. This equation, along with the equilibrium equations, allows for the calculation of the reactions at the supports.

In summary, statically indeterminate systems require a combination of equilibrium equations and compatibility conditions for their analysis. The compatibility conditions are derived from the physical and geometric constraints of the system, ensuring a continuous and consistent deformation pattern. Through the systematic application of these principles, engineers can solve statically indeterminate systems, providing critical insights into the behavior of complex structures under various loading conditions.

Chapter 9: Material Properties & Processing

Material Properties Overview

Understanding the multifaceted nature of material properties is crucial for engineers to design, analyze, and improve mechanical systems and components effectively. These properties, categorized into chemical, electrical, mechanical, physical, and thermal, provide a comprehensive framework for evaluating materials under various conditions and applications.

Chemical properties encompass the material's behavior when it undergoes a chemical change or reaction. Corrosion resistance, a critical chemical property, is vital in selecting materials for environments prone to chemical exposure. For instance, stainless steel's chromium content forms a protective layer, enhancing its corrosion resistance, crucial for applications in chemical processing industries.

Electrical properties include conductivity, resistivity, and dielectric strength. Materials with high electrical conductivity, such as copper and aluminum, are preferred for electrical wiring and components. Conversely, insulators like rubber and glass, characterized by their high resistivity and dielectric strength, are essential for preventing unwanted electrical transmission, ensuring safety and functionality in electrical systems.

Mechanical properties determine a material's response to mechanical forces. Tensile strength, ductility, hardness, and toughness are paramount in assessing a material's suitability for structural applications. For example, the tensile strength of a material, defined by the maximum stress it can withstand while being stretched or pulled before breaking, is a key factor in selecting materials for load-bearing structures.

Physical properties include density, melting point, and thermal expansion. These properties affect material selection based on weight considerations, operational temperature ranges, and dimensional stability. Lightweight materials with low density, such as aluminum, are favored in aerospace applications for fuel efficiency, while materials with low thermal expansion coefficients, like Invar, are used in applications requiring high dimensional stability across temperature changes.

Thermal properties such as specific heat capacity, thermal conductivity, and thermal diffusivity play a significant role in materials' behavior under thermal loads. Materials with high thermal conductivity, like copper, efficiently transfer heat and are used in heat sinks and heat exchangers. Conversely, materials with high specific heat capacity, like water, are effective in thermal energy storage, highlighting the importance of understanding thermal properties in energy management and conservation strategies.

Incorporating these material properties into the engineering design process enables the development of more efficient, reliable, and sustainable mechanical systems and components. By systematically analyzing and selecting materials based on their chemical, electrical, mechanical, physical, and thermal properties, engineers can optimize performance, durability, and cost-effectiveness, addressing the complex challenges of modern engineering projects.

Stress-Strain Diagrams

The stress-strain diagram is a fundamental tool in understanding the mechanical behavior of materials under load. When a material is subjected to a force, it deforms, and the relationship between the applied stress σ and the resulting strain ϵ provides critical insights into its strength and elasticity. Stress, defined as force per unit area $\left(\frac{F}{A}\right)$, measures the internal forces within a material, while strain, the ratio of deformation over original length $\left(\frac{\Delta L}{L_0}\right)$, quantifies the deformation extent.

Elastic Region: Initially, most materials exhibit a linear relationship between stress and strain, known as Hooke's Law, where $\sigma = E\epsilon$. In this region, deformation is elastic, meaning the material will return to its original shape upon load removal. The slope of the linear portion of the stress-strain curve represents the modulus of elasticity E, a measure of the material's stiffness.

Yield Point: As the stress increases, a point is reached where the material begins to deform plastically. The yield point marks the end of elastic behavior and the start of permanent deformation. For some materials, this transition is sharp and easily identified, while for others, it is more gradual, necessitating a 0.2% offset method to define a yield strength.

Ultimate Strength: With further strain, the material reaches its maximum strength, known as the ultimate tensile strength (UTS). Beyond this point, the material will start to neck and weaken until fracture.

Fracture: The stress decreases after the ultimate strength point, leading to material failure or fracture. The area under the stress-strain curve up to the fracture point represents the material's toughness, indicating its ability to absorb energy before failing.

Ductility and Brittleness: The extent of strain at fracture provides insight into the material's ductility or brittleness. Ductile materials, like many metals, undergo significant plastic deformation before breaking, showing a large area under their stress-strain curve. Brittle materials, such as ceramics and glass, fracture with minimal plastic deformation, indicated by a relatively small area under the curve.

Engineering Stress-Strain vs. True Stress-Strain: It is important to distinguish between engineering stress-strain and true stress-strain diagrams. Engineering stress and strain are based on the original dimensions of the material, whereas true stress and strain account for the continuous change in dimensions during deformation. True stress is higher than engineering stress beyond the yield point, as it considers the actual cross-sectional area of the necking material.

Materials exhibit different stress-strain behaviors based on their composition, processing, and temperature. Metals, polymers, ceramics, and composites each have unique diagrams reflecting their mechanical properties. For instance, polymers may show a nonlinear elastic region due to their molecular structure, while metals typically exhibit a clear elastic and plastic region due to dislocation movements.

Grasping the stress-strain diagram of a material is essential for engineers to anticipate how materials will respond to various loading conditions, which aids in the design of components capable of enduring operational stresses without failure. This expertise is vital for selecting suitable materials for specific applications, taking into account attributes such as strength, ductility, toughness, and elasticity to ensure reliability and safety in engineering designs.

Ferrous Metals: Properties and Uses

Ferrous metals, primarily composed of iron, are pivotal in the engineering and construction sectors due to their distinctive properties, widespread availability, and versatility in applications. **Steel** and **cast iron** are the two most prominent categories of ferrous metals, each possessing unique characteristics that make them suitable for various engineering applications. Steel, an alloy of iron and carbon, stands out for its exceptional strength-to-weight ratio, making it an indispensable material in the construction of buildings, bridges, and vehicles. The addition of other elements such as chromium, nickel, and molybdenum can further enhance its properties, leading to the creation of stainless steel variants known for their corrosion resistance. This attribute is particularly beneficial in environments prone to corrosive elements, ensuring longevity and durability of structures.

Cast iron, on the other hand, is recognized for its excellent castability and wear resistance. With a higher carbon content than steel, cast iron exhibits a lower melting point, which facilitates its casting into intricate shapes. This property is highly valued in the manufacturing of engine blocks, pipes, and cookware. However, cast iron's brittleness, a result of its high carbon content, limits its application in scenarios requiring ductility and toughness.

The processing of ferrous metals involves several stages, starting from the extraction of iron ore, followed by its refinement to produce pig iron. The **Bessemer Process**, one of the earliest methods for converting pig iron into steel, involves blowing air through molten pig iron to oxidize excess carbon, resulting in a more malleable steel. Modern steelmaking has evolved to include processes such as the **Basic Oxygen Steelmaking (BOS)** and **Electric Arc Furnace (EAF)** methods, which offer improved efficiency, control, and environmental benefits. The BOS method, for instance, uses oxygen to reduce carbon content in pig iron, while the EAF method recycles scrap steel using high-power electric arcs, highlighting the industry's shift towards sustainability.

Heat treatment processes such as annealing, quenching, and tempering are crucial in modifying the mechanical properties of ferrous metals to meet specific application requirements. **Annealing** involves heating the metal to a specific temperature and then slowly cooling it to make it more ductile and reduce internal stresses. **Quenching**, a process of rapid cooling, increases the hardness and strength of the metal. **Tempering**, which follows quenching, aims to achieve a balance between hardness and ductility by reheating the metal to a temperature below its critical point.

The surface treatment of ferrous metals also plays a significant role in enhancing their performance and longevity. Techniques such as galvanization, which involves coating steel with a layer of zinc, protect against corrosion. Similarly, carburizing, a process of adding carbon to the surface of low-carbon steel, improves wear resistance, making it suitable for high-stress applications like gears and bearings.

In conclusion, the properties, uses, and processing of ferrous metals such as steel and cast iron are foundational to their widespread application in engineering and construction. Understanding these aspects allows engineers to select and manipulate materials effectively, ensuring the structural integrity, durability, and performance of their projects.

Nonferrous Metals: Properties & Applications

Nonferrous metals, notably aluminum and copper, play pivotal roles in engineering applications due to their distinct properties and versatility. Aluminum, characterized by its low density, high corrosion resistance, and excellent conductivity, finds extensive use in aerospace, automotive, and construction industries. Its lightweight nature contributes significantly to fuel efficiency in transportation applications, while its formability and corrosion resistance make it an ideal choice for structural components and packaging materials. The thermal and electrical conductivity of aluminum, although lower than copper, is sufficiently high for it to be widely used in electrical transmission lines and heat exchangers.

Copper, on the other hand, stands out for its superior electrical and thermal conductivity, making it indispensable in electrical wiring, electronics, and motor manufacturing. Its excellent ductility allows it to be drawn into thin wires, while its thermal conductivity is essential for cooling systems and heat sinks. Additionally, copper's antimicrobial properties have led to its use in medical and architectural applications, where reducing the spread of bacteria is crucial.

The processing of these metals involves various methods to alter their physical and mechanical properties to suit specific applications. For aluminum, the Bayer process for extracting alumina from bauxite ore, followed by the Hall-Héroult process for smelting alumina to produce aluminum, is widely adopted. Post-extraction, aluminum can undergo work hardening, heat treatment, and alloying with elements such as silicon, magnesium, and zinc to enhance its strength, ductility, and corrosion resistance.

Copper extraction primarily involves pyrometallurgy, with the most common method being smelting of sulfide ores to produce copper matte, followed by electrorefining to achieve high purity levels. Like aluminum, copper can be alloyed, most notably with zinc to produce brass and with tin to produce bronze, each alloy offering a unique combination of properties for specific applications.

The sustainability aspects of nonferrous metals, particularly recycling, are significant. Both aluminum and copper boast high recyclability without loss of properties, contributing to energy savings and reduced environmental impact compared to primary production. The recycling of aluminum requires only about 5% of the energy used to produce it from ore, while recycled copper saves up to 85% of the energy used in its primary production.

In engineering design, the selection of nonferrous metals is guided by considerations of mechanical properties, such as strength, hardness, and fatigue resistance, as well as physical properties like density, conductivity, and thermal expansion. Engineers must also consider environmental factors, including corrosion resistance and material availability, to ensure the sustainability and longevity of their designs.

The application of nonferrous metals in engineering is further enhanced by surface treatments and coatings that improve their performance and durability. Anodizing, for example, can increase the corrosion resistance and wear properties of aluminum, while plating and cladding can protect copper surfaces and enhance their appearance.

Engineered Materials Overview

Engineered materials, including **composites**, **polymers**, and other synthetically developed substances, have revolutionized the field of material science and engineering by offering properties that surpass those of conventional materials. These materials are meticulously designed to meet specific requirements of strength, durability, weight reduction, and cost-effectiveness, making them indispensable in modern engineering applications.

Composites are materials made from two or more constituent materials with significantly different physical or chemical properties, that when combined, produce a material with characteristics different from the individual components. The most common example is **fiberglass**, which combines glass fibers with a polymer matrix to create a material that is strong, durable, and lightweight. Another example is **carbon fiber reinforced polymers (CFRPs)**, known for their exceptional strength-to-weight ratio, making them ideal for aerospace, automotive, and sports equipment. The key to the success of composites lies in their ability to be engineered for specific applications, allowing for the optimization of properties such as tensile strength, stiffness, and thermal resistance.

Polymers, on the other hand, are long chains of molecules that provide a wide range of properties based on their composition and structure. **Thermoplastics**, such as polyethylene (PE) and polyvinyl chloride (PVC), can be melted and reshaped multiple times without significant degradation, making them highly versatile for manufacturing processes like injection molding and extrusion. **Thermosetting plastics**, such as epoxy resins, undergo a chemical reaction during processing that sets their shape permanently, offering superior mechanical strength and thermal stability. Polymers can be engineered to exhibit specific characteristics, including high impact resistance, flexibility, or electrical insulation, making them suitable for a vast array of applications from packaging to electrical components.

Engineered materials also include **advanced ceramics** and **metal matrix composites (MMCs)**. Advanced ceramics, such as silicon nitride and zirconia, offer high temperature resistance, hardness, and wear resistance, ideal for applications in cutting tools, engine components, and medical devices. MMCs combine metal matrices with ceramic or other metallic reinforcements to enhance strength, wear resistance, and high-temperature performance, finding applications in automotive, aerospace, and military sectors.

The development and processing of engineered materials require a deep understanding of material science, chemistry, and engineering principles. Techniques such as **layered manufacturing**, **3D printing**, and **nanotechnology** are employed to create materials with nano-scale precision, enabling the development of materials with unprecedented properties. For instance, **nanocomposites** incorporate nanoparticles into polymers or metals to improve mechanical properties, electrical conductivity, and thermal stability.

The selection of engineered materials for a specific application involves considering factors such as mechanical properties, environmental resistance, manufacturing processes, and cost. Engineers must evaluate the trade-offs between material performance and expense to determine the most suitable material solution that meets the design criteria and functional requirements of the component or system.

Manufacturing Processes

Manufacturing processes are fundamental to the engineering and production of mechanical components and systems. These processes can be broadly categorized into casting, machining, and forming techniques, each with its unique set of principles, applications, and considerations for material selection and design. **Casting** is a manufacturing process where liquid material is poured into a mold, which contains a hollow cavity of the desired shape, and then allowed to solidify. The solidified part, known as the casting, is ejected or broken out of the mold to complete the process. This method is advantageous for creating complex shapes that would be difficult or uneconomical to make by other methods. Variations of casting include sand casting, die casting, and investment casting, each suitable for different materials and precision levels.

Machining involves the removal of material from a workpiece, shaping it to the desired form and dimensions. Common machining operations include turning, milling, drilling, and grinding. Turning operations are carried out on a lathe, where the workpiece rotates against a cutting tool to remove material. Milling, on the other hand, involves a rotating tool removing material from a stationary workpiece. Drilling creates round holes in a workpiece, typically using a rotating drill bit, while grinding uses an abrasive wheel to finish the surface of the workpiece to a high degree of precision and surface quality. Machining is critical for producing parts that require tight tolerances and smooth finishes.

Forming processes shape materials through deformation without adding or removing material. This category includes a wide range of techniques such as forging, where metal is shaped by localized compressive forces; stamping, where sheets of metal are pressed into the desired shape; and bending, where force is applied to produce a bend in the material. Forming is often chosen for its efficiency in mass production and the ability to create parts with enhanced mechanical properties due to work hardening.

Each of these manufacturing processes requires careful consideration of the material properties, desired part geometry, tolerances, surface finish, and production volume to select the most appropriate and cost-effective method. Understanding the capabilities and limitations of casting, machining, and forming is essential for engineers to design parts that can be manufactured efficiently and meet the required specifications.

Phase Diagrams and Heat Treatment

Phase diagrams serve as fundamental tools in materials science, providing a graphical representation of the states of matter of a substance as a function of temperature, pressure, and composition. These diagrams are pivotal for understanding the conditions under which a material changes phase, such as from solid to liquid or liquid to gas, and the impact of these changes on material properties. In the context of engineering materials, particularly metals and alloys, phase diagrams are essential for predicting the behavior of materials under various thermal and mechanical conditions, enabling engineers to tailor material properties to specific applications through controlled heating and cooling processes, known as heat treatment.

The equilibrium phase diagram typically displays temperature on the vertical axis and composition on the horizontal axis, delineating regions where different phases exist or coexist. For example, in a binary alloy system, the diagram might show regions of solid solution, mixtures of solid solutions, and liquid phases. Key features of phase diagrams include lines or curves, such as the liquidus, solidus, and solvus, which define the boundaries between different phases. The eutectic point, where a mixture of phases coexists at a specific composition and temperature, is of particular interest for its implications in material processing and properties.

Heat treatment processes, such as annealing, quenching, and tempering, are directly informed by phase diagrams. Annealing, for instance, involves heating a material to a specific temperature within a phase region and then cooling it at a controlled rate to alter its microstructure, thereby relieving internal stresses, increasing ductility, and reducing hardness. Quenching, on the other hand, involves rapid cooling from a high temperature, typically in the austenitic phase region for steels, to trap a high-temperature phase at room temperature, significantly increasing hardness and strength. Subsequent tempering adjusts the properties of quenched materials by reheating to a temperature below the eutectic, allowing for controlled precipitation and growth of secondary phases, which improves toughness.

The manipulation of phase transformations through heat treatment enables the engineering of materials with tailored mechanical properties, such as strength, toughness, and ductility, essential for specific applications. For instance, the heat treatment of steel can produce a wide range of properties, from the high strength and low ductility of martensitic steels to the balanced strength and ductility of tempered steels. Understanding the phase diagram of an alloy system allows engineers to predict the outcome of heat treatments and design processes that achieve the desired material characteristics.

Materials Selection Criteria

In the realm of engineering design, the selection of materials is a critical step that directly influences the performance, durability, and overall success of the final product. This process requires a meticulous evaluation of the material properties to ensure they align with the specific requirements of the application. The criteria for materials selection encompass a broad spectrum of physical, chemical, mechanical, and economic factors, each playing a pivotal role in the decision-making process.

Physical properties are fundamental to understanding how a material will behave under various conditions. Density, for instance, affects the weight and mass of the final product, which is crucial for automotive and aerospace applications where weight reduction is a priority. Thermal conductivity and expansion are also critical for applications subjected to fluctuations in temperature, ensuring that materials can withstand or adapt to thermal stresses without failure.

Chemical properties focus on a material's resistance to corrosion, oxidation, and chemical degradation. This is particularly important in harsh environments where exposure to corrosive substances or extreme conditions can lead to material failure. Selecting materials with appropriate chemical stability ensures longevity and reliability of the component in its intended environment.

Mechanical properties include strength, ductility, hardness, and toughness. These characteristics determine a material's ability to withstand mechanical stresses without deforming or breaking. For structural applications, high strength and toughness are desirable to resist impact and wear, while ductility is important for components that must undergo deformation during their service life without fracturing.

Economic factors play a significant role in materials selection, as cost-effectiveness is often a primary concern in engineering projects. This includes not only the initial cost of the material but also considerations related to manufacturing processes, lifecycle costs, maintenance, and recyclability. A material that is initially more expensive but offers superior performance or durability may prove more cost-effective in the long term.

The selection process often involves trade-offs, as it is rare for a single material to excel in all desired properties. Engineers must prioritize the criteria based on the application's specific needs, using a systematic approach to evaluate and compare potential materials. Tools such as decision matrices or software for materials selection can aid in this process by quantifying the importance of each criterion and providing a comparative analysis of how well each material meets the requirements.

In addition to these primary criteria, engineers must also consider **manufacturability** and **compatibility** with other materials. Some materials may require specialized processing techniques that can increase production costs or complexity. Similarly, when multiple materials are used together, their compatibility must be assessed to prevent issues such as galvanic corrosion or differential thermal expansion that could compromise the integrity of the assembly.

Ultimately, the selection of materials is a complex and critical component of the engineering design process, requiring a deep understanding of material properties and a strategic approach to balancing performance, cost, and manufacturability. By carefully considering these factors, engineers can make informed decisions that enhance the functionality, durability, and economic viability of their designs, ensuring that the final product meets or exceeds the expectations of its intended application.

Corrosion Mechanisms and Control

Corrosion, a pervasive issue affecting materials, particularly metals, in engineering, is the process by which materials deteriorate due to interactions with their environment. Understanding the mechanisms of corrosion is crucial for engineers to design and implement effective control strategies, ensuring the longevity and reliability of components and structures. The primary types of corrosion include uniform or general corrosion, galvanic corrosion, crevice corrosion, pitting corrosion, intergranular corrosion, stress-corrosion cracking, and erosion-corrosion. Each type presents unique challenges and requires specific prevention and control measures.

Uniform or General Corrosion occurs evenly across the surface of a material. It is predictable, with metal loss occurring at a steady rate over time. Control strategies for uniform corrosion involve selecting materials with inherent resistance to corrosion in the given environment, applying protective coatings, or using cathodic protection systems to mitigate the electrochemical reactions leading to corrosion.

Galvanic Corrosion arises when two dissimilar metals are in electrical contact in the presence of an electrolyte, leading to accelerated corrosion of the more anodic material. Prevention involves isolating metals from direct contact, either by using non-conductive barriers or coatings, or by selecting metals close together in the galvanic series to minimize potential differences.

Crevice Corrosion is localized corrosion occurring in confined spaces or crevices, where the electrolyte becomes stagnant, leading to a differential aeration cell. Designing to avoid crevices, using sealants, and employing materials resistant to crevice corrosion are effective strategies for control.

Pitting Corrosion is a form of extremely localized attack that leads to the creation of small holes or pits. This type of corrosion is particularly insidious as it can lead to failure with minimal overall material loss. Control measures include the use of pitting-resistant materials, maintaining clean surfaces to prevent initiation, and employing inhibitors to slow the corrosion process.

Intergranular Corrosion occurs along the grain boundaries of a material, often where precipitates form, creating paths for corrosion that can lead to material failure. Solution treatment, which dissolves precipitates, followed by rapid quenching, and the use of low-carbon and stabilized grades of stainless steel are effective in preventing intergranular corrosion.

Stress-Corrosion Cracking (SCC) is the growth of crack formation in a corrosive environment. It can be catastrophic as it combines the effects of stress and corrosion. Control strategies include reducing residual stresses through stress relief treatments, using materials resistant to SCC for the given environment, and minimizing the corrosive aspects of the environment.

Erosion-Corrosion results from the combined action of corrosion and the mechanical wear of a material surface due to fluid flow. The control of erosion-corrosion involves reducing fluid velocities, using erosion-resistant materials or coatings, and streamlining designs to minimize turbulent flow.

General practices for corrosion control include regular maintenance and inspection, environmental modification to reduce corrosive factors, and the use of corrosion inhibitors. The selection of appropriate materials and design considerations play a pivotal role in minimizing the risk of corrosion. Engineers must also consider the economic impact of corrosion control measures, balancing the cost of prevention with the potential costs of corrosion damage and failure. Effective corrosion control is a multifaceted challenge that requires a comprehensive understanding of both the materials involved and the environmental conditions to which they are exposed.

Failure Mechanisms and Prevention

Material failure mechanisms are critical considerations in the design and analysis of engineering components and systems. Understanding these failure modes—thermal failure, fatigue, fracture, and creep—is essential for predicting how materials behave under various conditions and for developing strategies to prevent such failures. Each of these failure mechanisms has distinct characteristics and is influenced by different factors, making it crucial for engineers to grasp their fundamentals and mitigation techniques.

Thermal Failure occurs when a material cannot withstand changes in temperature. It may result from excessive thermal stresses due to differential expansion or contraction, leading to cracking or other forms of degradation. To prevent thermal failure, materials with appropriate thermal expansion coefficients and high thermal shock resistance should be selected. Additionally, design strategies such as including expansion joints or using thermal barriers can mitigate the risks associated with thermal stresses.

Fatigue is a failure mechanism that results from cyclic loading, leading to the initiation and propagation of cracks over time. Even stresses that are below the material's yield strength can cause fatigue failure if applied repeatedly. Preventing fatigue involves reducing stress concentrations by optimizing design geometries, selecting materials with high fatigue strength, and applying surface treatments that improve resistance to crack initiation. Regular inspection and maintenance schedules can also detect early signs of fatigue, allowing for corrective actions before catastrophic failure occurs.

Fracture refers to the propagation of cracks through a material, leading to its separation into two or more pieces. Fracture can be brittle or ductile, with brittle fracture occurring suddenly and without significant deformation, while ductile fracture is preceded by noticeable plastic deformation. To prevent fracture, materials with adequate toughness should be selected, particularly for applications involving impact or shock loading. Design considerations, such as avoiding sharp corners and notches, can also reduce stress concentrations that may initiate cracks.

Creep is the time-dependent deformation of materials under constant stress, typically at high temperatures relative to the material's melting point. Creep can lead to excessive deformation and eventual failure. Preventing creep involves selecting materials with high creep resistance for applications at elevated temperatures and designing components to operate below stress and temperature thresholds that accelerate creep deformation. Additionally, using alloys specifically designed to resist creep, such as those strengthened by precipitation hardening, can significantly enhance the longevity of components in high-temperature environments.

Preventing material failure necessitates a thorough grasp of the mechanisms involved along with a strategic focus on material selection, component design, and operational management. Considering the specific conditions that materials will encounter and implementing suitable mitigation strategies allows engineers to significantly improve the durability and reliability of engineering systems, thus ensuring safety and performance throughout their intended lifespan.

Chapter 10: Fluid Mechanics

Fluid Properties

Understanding fluid properties is crucial for analyzing and predicting fluid behavior under various conditions, which is essential for the FE Mechanical exam and practical engineering applications. **Density** (ρ), a fundamental property, represents the mass per unit volume of a fluid and is a critical parameter in the equations of fluid statics and dynamics. It influences buoyancy, pressure, and flow characteristics. The density of a fluid can vary with temperature and pressure, a dependency that must be accounted for in calculations involving gases and compressible flows. For incompressible flows, such as most liquid systems, density is typically considered constant.

Viscosity, another key fluid property, measures a fluid's resistance to flow and shear. It is categorized into dynamic viscosity (μ) and kinematic viscosity (ν), where kinematic viscosity is the dynamic viscosity divided by the density of the fluid ($\nu = \mu/\rho$). The viscosity of a fluid affects the laminar or turbulent nature of the flow, the formation of boundary layers, and the energy losses due to friction within the fluid. Engineers must understand how temperature and pressure affect viscosity to accurately predict flow behavior in real-world scenarios.

Surface tension (γ) is the force per unit length acting at the interface between a liquid and a gas, or between two immiscible liquids, due to molecular attractions. Surface tension is responsible for phenomena such as the formation of droplets, capillary rise, and the shape of the liquid surfaces. In engineering, surface tension impacts the behavior of small-scale flows, influences heat and mass transfer processes, and affects the design and analysis of devices dealing with fluid interfaces.

Each of these properties plays a vital role in fluid mechanics, influencing the analysis and design of engineering systems, from hydraulic pumps and aircraft wings to chemical reactors and HVAC systems. Engineers leverage these properties to solve problems related to fluid flow, energy transfer, and material selection, ensuring the efficiency, safety, and effectiveness of mechanical systems. Understanding the interplay between density, viscosity, and surface tension allows engineers to model and predict fluid behavior under various conditions, a skill that is indispensable for passing the FE Mechanical exam and succeeding in professional engineering practice.

Fluid Statics and Hydrostatic Forces

In the realm of fluid mechanics, understanding the principles of fluid statics is crucial for engineers preparing for the FE Mechanical exam. This section delves into the pressure distribution within static fluids and the hydrostatic forces exerted on submerged surfaces, which are foundational concepts in fluid mechanics. The pressure at any point within a static fluid is isotropic, meaning it acts equally in all directions. This characteristic is pivotal in analyzing the forces on submerged surfaces, whether they are plane or curved.

The pressure in a fluid at rest increases with depth due to the weight of the fluid above. This relationship is quantified by the hydrostatic pressure equation, $P = P_0 + \rho g h$, where P is the pressure at a certain depth, P_0 is the atmospheric pressure, ρ is the fluid density, g is the acceleration due to gravity, and h is the depth below the surface. This equation is instrumental in calculating the pressure exerted by the fluid at any point and is a fundamental principle that engineers must master.

When analyzing hydrostatic forces on surfaces, it is essential to consider the shape and orientation of the surface. For a horizontal plane surface submerged in a fluid, the force exerted by the fluid is directly proportional to the depth of the centroid of the area below the fluid surface and the area itself. The force can be calculated using the formula $F = \rho g h A$, where A is the area of the surface. This force acts vertically upwards at the center of pressure, which, for horizontal surfaces, coincides with the centroid of the area.

For vertical or inclined plane surfaces submerged in a fluid, the calculation of hydrostatic forces becomes slightly more complex due to the variation of pressure over the surface. The total hydrostatic force is still determined by the product of the fluid's weight above the centroid of the area and the area itself, but the point of action, or the center of pressure, does not coincide with the centroid since the pressure varies with depth.

The concept of buoyancy, governed by Archimedes' principle, states that the buoyant force on a submerged object is equal to the weight of the fluid displaced by the object. This principle is vital for understanding the stability of floating and submerged bodies in static fluids. Engineers use this principle to analyze and design various structures and devices, ensuring they perform as intended when interacting with fluids.

Energy, Impulse, and Momentum Principles

In fluid mechanics, the principles of energy, impulse, and momentum are foundational for understanding how fluid systems behave under various conditions. These principles are governed by the laws of conservation, which state that, in an isolated system, these quantities remain constant over time. This section delves into the application of these conservation principles within fluid systems, providing a detailed examination of their implications for the FE Mechanical exam.

Energy in fluid systems is primarily considered in terms of mechanical energy, which can be in the form of kinetic energy, potential energy, or internal energy. The Bernoulli equation is a manifestation of the conservation of energy principle in fluid flow, expressed as $P + \frac{1}{2}\rho v^2 + \rho g h = \text{constant}$, where P represents the pressure energy, $\frac{1}{2}\rho v^2$ the kinetic energy per unit volume, and $\rho g h$ the potential energy per unit volume. This equation is pivotal in analyzing fluid flow in pipes, channels, and around objects, providing insights into how energy is converted and conserved within a fluid stream.

Impulse reflects the change in momentum of a fluid system resulting from external forces. The impulse-momentum equation, $\vec{F}\Delta t = \Delta(m\vec{v})$, where $\vec{F}$ is the force applied, Δt the time interval, m the mass of the fluid, and $\vec{v}$ its velocity, is crucial for understanding how forces affect fluid motion. This principle is applied in analyzing fluid jets, forces on pipe bends, and the impact of fluid flow on structures, enabling engineers to predict and mitigate potential issues arising from fluid dynamics.

Momentum conservation in fluid mechanics is often discussed in the context of the Navier-Stokes equations, which describe the motion of viscous fluid substances. However, a more simplified approach for the FE Mechanical exam focuses on the steady, incompressible flow of an ideal fluid, where the principle of conservation of linear momentum, $\sum \vec{F} = \frac{d\vec{p}}{dt}$, with $\vec{p}$ being the momentum of the fluid and $\sum \vec{F}$ the sum of external forces, is used to analyze forces in fluid systems such as pumps, turbines, and nozzles. This analysis is essential for designing and optimizing fluid systems to ensure efficient operation and safety.

Understanding and applying these principles require a thorough grasp of the underlying physics and the ability to perform quantitative analyses. For engineers preparing for the FE Mechanical exam, mastering these concepts is not only crucial for passing the exam but also for their future careers in designing and analyzing fluid systems in various engineering applications. The ability to predict how energy, impulse, and momentum will be conserved or transformed in fluid systems underpins much of the work in mechanical engineering, from HVAC systems to hydraulic machinery, making this knowledge indispensable.

Internal Flow in Pipes and Ducts

In the realm of fluid mechanics, the analysis of internal flow within pipes and ducts is paramount for engineers aiming to excel in the FE Mechanical exam. This section delves into the critical aspects of fluid flow, emphasizing friction losses and flow rate analysis, which are essential for designing efficient fluid transport systems. Understanding the behavior of fluids in confined spaces requires a comprehensive grasp of the principles governing fluid dynamics, including the Reynolds number, which characterizes the flow regime as laminar or turbulent.

Friction losses in pipes are predominantly attributed to the roughness of the pipe's interior surface and the flow velocity. The Darcy-Weisbach equation, $h_f = f\left(\frac{L}{D}\right)\left(\frac{v^2}{2g}\right)$, where h_f represents the frictional head loss, f the friction factor, L the length of the pipe, D its diameter, v the velocity of the fluid, and g the acceleration due to gravity, serves as a fundamental formula for calculating head loss due to friction. The friction factor, f, varies with the flow regime and the pipe's roughness, necessitating the use of the Moody chart or the Colebrook-White equation for its determination in turbulent flow conditions.

Flow rate analysis in pipes and ducts hinges on the continuity equation and Bernoulli's principle. The continuity equation, $A_1 v_1 = A_2 v_2$, where A represents the cross-sectional area of the pipe and v the velocity of the fluid, underscores the principle of mass conservation in fluid flow. Bernoulli's equation, $P_1 + \frac{1}{2}\rho v_1^2 + \rho g h_1 = P_2 + \frac{1}{2}\rho v_2^2 + \rho g h_2$, where P denotes the pressure, ρ the fluid density, and h the elevation head, elucidates the energy conservation in a flowing fluid. This equation is instrumental in analyzing flow rates and pressure drops, especially in systems where elevation changes and pumping are involved.

The practical application of these principles extends to the design and analysis of piping systems for water distribution, HVAC, and industrial processes. Engineers must adeptly apply these concepts to ensure that fluid transport systems are both efficient and meet the required performance specifications. This entails not only a thorough understanding of fluid dynamics but also an ability to perform complex calculations that factor in the various parameters affecting flow within pipes and ducts.

Moreover, the selection of pipe material and diameter is a critical decision in the design process, influenced by the fluid properties, required flow rate, and economic considerations. The engineer's ability to optimize these parameters for minimal energy consumption and cost while ensuring system reliability and compliance with standards is a testament to their proficiency in fluid mechanics.

In preparing for the FE Mechanical exam, engineers must not only familiarize themselves with the theoretical aspects of internal flow but also develop the analytical skills necessary to solve real-world engineering problems. This includes the ability to interpret and apply empirical data, utilize computational tools for flow analysis, and understand the implications of fluid dynamics principles in the context of modern engineering practices.

External Flow: Boundary Layers and Drag

In the study of fluid mechanics, external flow plays a crucial role, particularly when analyzing how fluids move over objects, which is a common scenario in mechanical engineering applications. This section delves into the dynamics of boundary layer formation and the resultant drag forces, both of which are pivotal for engineers preparing for the FE Mechanical exam. The boundary layer is a thin layer of fluid that forms on the surface of an object as it moves through a fluid or as fluid moves past a stationary object. Within this layer, fluid velocity increases from zero at the surface (due to the no-slip condition) to approximately the free stream velocity at the outer edge of the boundary layer. The development of the boundary layer is influenced by the viscosity of the fluid and the shape of the object, leading to either laminar or turbulent flow.

The transition from laminar to turbulent flow within the boundary layer is characterized by the Reynolds number, a dimensionless quantity that depends on the velocity of the fluid, the characteristic length (such as the chord length of an airfoil), and the kinematic viscosity of the fluid. For lower Reynolds numbers, the flow remains laminar, exhibiting smooth streamlines and predictable behavior. As the Reynolds number increases, the flow becomes unstable and transitions to a turbulent state, marked by chaotic fluid motion and mixing. This transition significantly affects the drag forces experienced by the object.

Drag forces are primarily composed of friction drag (skin friction) and pressure drag (form drag). Friction drag arises from the shear stress between the fluid and the object's surface, predominantly affecting objects with a large surface area parallel to the flow direction. Pressure drag results from the pressure differential between the front and back of the object, significantly influenced by the object's shape. Streamlined shapes reduce pressure drag by allowing the fluid to flow smoothly around them, minimizing the wake region behind the object where low pressure can develop.

The analysis of drag forces is essential for optimizing the design of various engineering systems, such as vehicles, aircraft, and wind turbines, to improve their efficiency and performance. Engineers use principles of external flow to predict and mitigate adverse effects of drag, employing strategies such as streamlining shapes, introducing boundary layer control methods (like vortex generators), and selecting materials with favorable surface characteristics.

A comprehensive understanding of boundary layer formation and drag forces necessitates a focus on the principles of fluid dynamics, supported by mathematical models and empirical data. Candidates preparing for the FE Mechanical exam must have a solid grasp of these concepts, as they form the basis for numerous questions related to fluid mechanics and are vital for the design and analysis of mechanical systems. Proficiency in external flow dynamics not only aids in successfully passing the exam but also provides future engineers with the essential knowledge to address practical challenges in fluid mechanics, contributing to the creation of efficient, high-performance engineering solutions.

Compressible Flow

Compressible flow, a critical aspect of fluid mechanics, becomes significant when dealing with flow velocities approaching or exceeding the speed of sound. At these high speeds, the density of the fluid can no longer be considered constant, and the flow behavior exhibits unique characteristics that are pivotal for engineers to understand, especially in the context of the FE Mechanical exam. The Mach number, a dimensionless quantity, is central to the study of compressible flow, defined as the ratio of the flow velocity to the local speed of sound, $M = \frac{v}{a}$, where v is the flow velocity and a is the speed of sound in the medium. The Mach number categorizes the flow regime into subsonic ($M < 1$), sonic ($M = 1$), supersonic ($M > 1$), and hypersonic ($M > 5$), each of which exhibits distinct physical phenomena and requires different analytical approaches.

Isentropic relations are fundamental in analyzing compressible flows, especially in adiabatic and reversible processes, where no heat is added or removed from the system. These relations provide a means to calculate changes in pressure, temperature, and density through the flow, assuming the process is isentropic. For instance, the relationship between pressure and density in an isentropic flow can be expressed as $P = K\rho^\gamma$, where P is the pressure, ρ is the density, γ is the specific heat ratio, and K is a constant. These equations are instrumental in designing and analyzing components like nozzles and diffusers, where compressible flow effects are significant.

Normal shocks represent a sudden and drastic change in the flow properties, such as velocity, temperature, and pressure, occurring when a supersonic flow decelerates to subsonic speeds across the shock wave. The analysis of normal shocks is crucial for understanding flow behavior in supersonic aircraft, wind tunnels, and various aerospace applications. The Rankine-Hugoniot relations describe the changes across a shock wave, providing equations to calculate the post-shock pressure, temperature, and density based on the pre-shock conditions and the Mach number. For example, the pressure ratio across a normal shock can be determined by $\frac{P_2}{P_1} = 1 + \frac{2\gamma}{\gamma + 1}(M_1^2 - 1)$, where P_1 and P_2 are the pre-shock and post-shock pressures, respectively, and M_1 is the Mach number of the flow approaching the shock.

Power and Efficiency in Fluid Systems

In fluid systems, **power** and **efficiency** are critical metrics that determine the performance of turbines and pumps, which are essential components in a wide range of engineering applications. Power in fluid systems is generally defined as the rate at which work is done or energy is transferred within the system. It can be expressed mathematically as $P = \dot{W}$, where P represents power and $\dot{W}$ denotes the rate of work done. For turbines, which convert fluid energy into mechanical work, the power output can be calculated using the formula $P_t = \dot{m} \cdot (h_1 - h_2)$, where $\dot{m}$ is the mass flow rate of the fluid through the turbine, and h_1 and h_2 are the specific enthalpies of the fluid at the inlet and outlet of the turbine, respectively. This equation highlights the importance of the enthalpy drop across the turbine in determining its power output.

Efficiency, on the other hand, is a measure of how effectively a turbine or pump converts energy from one form to another. The efficiency of a turbine is given by $\eta_t = \frac{\text{Actual Work Output}}{\text{Isentropic Work Output}}$, which can also be expressed as $\eta_t = \frac{h_1 - h_2}{h_1 - h_{2s}}$, where h_{2s} represents the specific enthalpy at the outlet of the turbine under isentropic conditions. This efficiency metric is crucial for evaluating the performance of turbines, as it accounts for the irreversibilities and losses that occur during the energy conversion process.

Pumps, conversely, require input power to increase the fluid's energy. The power input to a pump can be calculated using $P_p = \dot{m} \cdot (h_2 - h_1)$, where h_2 and h_1 are the specific enthalpies of the fluid at the pump outlet and inlet, respectively. The efficiency of a pump is determined by $\eta_p = \dfrac{\text{Isentropic Work Input}}{\text{Actual Work Input}}$, or equivalently, $\eta_p = \dfrac{h_{2s} - h_1}{h_2 - h_1}$, where h_{2s} is the specific enthalpy at the pump outlet under isentropic conditions. This efficiency measure is vital for assessing how effectively a pump converts mechanical work into fluid energy, highlighting the significance of minimizing energy losses to enhance pump performance.

Calculating power and efficiency in fluid systems is crucial for engineers involved in the design and analysis of turbines and pumps. These calculations enhance system performance and lead to the creation of energy-efficient and cost-effective engineering solutions. Focusing on improving the efficiency of turbines and pumps allows engineers to lower energy consumption and operational expenses, thereby promoting sustainability and environmental responsibility in engineering projects.

Performance Curves for Pumps and Fans

Performance curves are essential tools in the analysis and selection of pumps, fans, and compressors, providing critical insights into how these devices will perform under various operating conditions. These curves graphically represent the relationship between several key parameters, such as flow rate, head, power consumption, and efficiency, allowing engineers to predict the behavior of a fluid-moving device in a system.

For pumps, the performance curve plots the head (pressure increase) against the flow rate. The curve typically starts high at zero flow (dead-head) and decreases as the flow rate increases, illustrating the pump's maximum head capability and how it diminishes with increased flow. The operating point of the pump, where the system curve intersects the pump curve, indicates the actual flow rate and head at which the pump will operate. Understanding this relationship is crucial for selecting a pump that meets the system's requirements without excessive energy consumption or wear.

Fans and blowers have similar performance curves, where the static pressure is plotted against the flow rate. These curves help in selecting fans that can deliver the required airflow against the resistance of the system, which includes ductwork, filters, and other components. The fan efficiency curve, which often accompanies the pressure-flow rate curve, shows the efficiency at different points, guiding the selection process towards the most energy-efficient option for the required operating conditions.

Compressors, particularly those used in HVAC and refrigeration, have performance curves that plot the refrigeration capacity (or mass flow rate) against the suction and discharge pressures, with additional curves for power consumption and isentropic efficiency. These curves are vital for selecting compressors that can achieve the desired cooling capacity while operating efficiently within the system's pressure constraints.

When analyzing performance curves, several factors must be considered:
- **The Best Efficiency Point (BEP)**: This is the point on the curve where the device operates with maximum efficiency. Operating too far from the BEP can lead to increased energy consumption and wear.
- **NPSH Required (NPSHR)** for pumps: This indicates the minimum net positive suction head required to prevent cavitation, a condition where vapor bubbles form and collapse within the pump, causing damage.
- **Surge and Stonewall Points** for compressors: These are the limits of stable operation. Operating beyond these points can cause damage or operational failure.

In summary, performance curves are indispensable for the proper selection and analysis of pumps, fans, and compressors. They provide a graphical representation of how these devices will perform under various conditions, allowing engineers to make informed decisions that balance efficiency, capacity, and longevity. Understanding how to read and utilize these curves is fundamental for mechanical engineers involved in the design and operation of systems incorporating these fluid-moving devices.

Scaling Laws for Fans and Compressors

Scaling laws for fans, pumps, and compressors are fundamental principles that allow engineers to predict how these devices will perform when scaled to different sizes or operated under varying conditions. These laws are derived from the basic principles of fluid dynamics and thermodynamics, taking into account the geometric, kinematic, and dynamic similarities between different scales. Understanding these scaling laws is crucial for the design and optimization of fluid systems, ensuring that devices operate efficiently and effectively across a range of applications.

The performance of fans, pumps, and compressors can be characterized by several key parameters, including flow rate Q, head or pressure difference ΔP, power consumption P, and efficiency η. Scaling laws relate these parameters to the physical dimensions of the device, the speed of operation, and the properties of the fluid. The most common scaling laws used in engineering practice are based on the principles of dimensional analysis and similarity.

Geometric Similarity requires that all corresponding dimensions of the models and prototypes are in the same proportion. This ensures that both the model and the real device have the same shape, but they may differ in size.

Kinematic Similarity is achieved when the velocity fields of the fluid in the model and the prototype are geometrically similar. This implies that the flow patterns in both the scaled model and the actual device are identical, which is essential for accurate prediction of fluid behavior.

Dynamic Similarity involves the forces acting in the fluid and requires that the ratios of all relevant forces (inertial, viscous, gravitational, etc.) are the same in the model and the prototype. This similarity ensures that the scaling laws accurately predict how changes in size or operating conditions will affect the performance of the device.

The application of these principles leads to specific scaling laws for fans, pumps, and compressors. For example, the flow rate Q scales with the cube of the linear dimension D and the speed N, the head or pressure difference ΔP scales with the square of the speed and the square of the linear dimension, and the power consumption P scales with the fifth power of the linear dimension and the cube of the speed. These relationships can be expressed as:

$Q \propto D^3 N$

$\Delta P \propto D^2 N^2$

$P \propto D^5 N^3$

Efficiency η, on the other hand, is generally considered to be independent of scale, assuming geometric, kinematic, and dynamic similarity are maintained. However, in practical applications, efficiency can be affected by factors such as surface roughness and Reynolds number, which may vary with scale.

By applying these scaling laws, engineers can predict how a change in size or speed of a fan, pump, or compressor will affect its flow rate, pressure difference, and power consumption. This is invaluable in the design process, allowing for the optimization of these devices for specific applications without the need for extensive experimental testing at full scale. Additionally, understanding the impact of scaling on efficiency can help in selecting the most appropriate device size and operating conditions to meet performance requirements while minimizing energy consumption.

Chapter 11: Thermodynamics

Properties of Ideal Gases and Substances

Understanding the properties of ideal gases and pure substances is fundamental for thermodynamic calculations, which are pivotal in various engineering applications, including HVAC systems, engine design, and refrigeration technologies. Ideal gases are hypothetical gases that perfectly follow the Ideal Gas Law $PV = nRT$, where P stands for pressure, V for volume, n for moles of gas, R for the ideal gas constant, and T for temperature in Kelvin. This law provides a cornerstone for understanding how gases behave under different conditions of temperature and pressure, assuming no forces between molecules and that the volume occupied by the gas molecules is negligible compared to the container volume.

For pure substances, which are materials consisting of a single type of particle, phase changes are a critical area of study. Pure substances can exist in different phases—solid, liquid, and gas—and can transition between these phases at specific temperature and pressure conditions. The phase behavior of a pure substance is often represented on a phase diagram, which plots pressure against temperature. Key points on this diagram include the triple point, where all three phases coexist in equilibrium, and the critical point, beyond which the distinction between liquid and gas phases ceases to exist.

The specific heat capacities of substances, both in constant volume C_v and constant pressure C_p scenarios, are also crucial. These capacities represent the amount of heat required to change the temperature of a unit mass of the substance by one degree Celsius at constant volume or pressure, respectively. For ideal gases, these values are constants, but for real substances, they vary with temperature. The relationship between C_p and C_v is given by the equation $C_p - C_v = R$, which is derived from the first law of thermodynamics and the ideal gas law.

Another important concept is the enthalpy of a substance, which is a measure of the total energy of a system. Enthalpy is particularly useful in processes that occur at constant pressure, where the change in enthalpy ΔH equals the heat added to or removed from the system. For ideal gases, the change in enthalpy can be directly calculated using C_p and the change in temperature, ΔT.

The thermodynamic properties of substances are also influenced by their molecular structure. For example, the equation of state for real gases, such as the Van der Waals equation, accounts for the volume occupied by gas molecules and the intermolecular forces, providing a more accurate description of gas behavior under non-ideal conditions than the Ideal Gas Law.

Energy Transfers in Thermodynamics

In thermodynamics, energy transfers are fundamental processes that occur in any system involving heat and work. These transfers are pivotal in determining the state and evolution of a system, governed by the first and second laws of thermodynamics. The first law, also known as the law of energy conservation, states that energy cannot be created or destroyed in an isolated system, implying that the total energy of the system is constant. This principle can be mathematically represented as $\Delta U = Q - W$, where ΔU is the change in internal energy of the system, Q is the heat added to the system, and W is the work done by the system on its surroundings. The positive sign of Q indicates heat added to the system, while the positive sign of W denotes work done by the system, following the convention that energy inputs to the system are positive, and energy outputs from the system are negative.

Heat transfer, a mode of energy transfer between physical systems, depends on the temperature difference between the systems and the properties of the medium through which the heat is transferred. There are three primary mechanisms of heat transfer: conduction, convection, and radiation. Conduction occurs through direct contact, where kinetic energy is transferred from high to low-temperature regions. The rate of heat transfer by conduction can be described by Fourier's law, $\dot{Q} = -kA\frac{dT}{dx}$, where $\dot{Q}$ is the rate of heat transfer, k is the thermal conductivity of the material, A is the cross-sectional area through which heat is transferred, and $\frac{dT}{dx}$ is the temperature gradient. Convection involves the transfer of heat by the movement of fluids, which can be natural, driven by buoyancy forces, or forced, induced by external means such as a pump or fan. Radiation, the transfer of energy by electromagnetic waves, can occur in a vacuum and is described by the Stefan-Boltzmann law, $Q = \sigma A T^4$, where σ is the Stefan-Boltzmann constant, A is the area of the emitting surface, and T is the absolute temperature of the surface.

Work transfer in thermodynamic systems can be realized in various forms, including mechanical work, such as the expansion or compression of gases, and electrical work, such as the work done by an electric current driven by a potential difference. The work done during a quasi-static process, a process that proceeds in infinitesimally small steps to remain in thermodynamic equilibrium, can be expressed as $W = \int_{V_1}^{V_2} PdV$, where W is the work done, P is the pressure, and dV is the differential change in volume, with the limits of integration from initial volume V_1 to final volume V_2.

The second law of thermodynamics introduces the concept of entropy, S, a measure of the disorder or randomness in a system, stating that in any spontaneous process, the total entropy of the system and its surroundings always increases. This law highlights the directionality of processes and the concept of irreversibility, emphasizing that energy quality degrades over time, limiting the amount of work that can be extracted from a system. The change in entropy can be calculated for a reversible process as $\Delta S = \int \frac{dQ_{rev}}{T}$, where dQ_{rev} is the reversible heat transfer and T is the absolute temperature at which the transfer occurs.

Understanding the mechanisms of energy transfer and the laws governing these processes is crucial for engineers to design and analyze systems for energy conversion, HVAC, refrigeration, and power generation, ensuring efficient and sustainable use of resources.

Laws of Thermodynamics

The **Laws of Thermodynamics** are foundational principles that govern the behavior of energy in physical systems, providing a framework for understanding how energy is conserved and transformed. These laws are critical for engineers to master, as they underpin the analysis and design of a wide range of mechanical systems, from engines and refrigerators to power plants and HVAC systems.

The First Law of Thermodynamics, often referred to as the law of energy conservation, states that energy cannot be created or destroyed in an isolated system. Mathematically, it is expressed as $\Delta U = Q - W$, where ΔU represents the change in internal energy of the system, Q is the heat added to the system, and W is the work done by the system. This law highlights the convertibility of heat and work into each other while maintaining the total energy of the system. It serves as the basis for energy balance calculations in engineering tasks, emphasizing that all energy entering a system must either increase the system's internal energy, be stored, or leave the system as work or heat.

The Second Law of Thermodynamics introduces the concept of entropy, a measure of disorder or randomness within a system. It states that for any spontaneous process, the total entropy of a system and its surroundings always increases. This law can be formulated in several ways, one of which is through the Kelvin-Planck statement: it is impossible for any device that operates on a cycle to receive heat from a single reservoir and produce a net amount of work. This principle underlies the inherent inefficiency of all real processes and machines, dictating that some energy is always lost to entropy, often as waste heat. The increase in entropy also implies the directionality of processes, indicating that while energy can be transformed from one form to another, the quality of energy degrades over time.

The Third Law of Thermodynamics asserts that as the temperature of a system approaches absolute zero, the entropy of the system approaches a minimum value. This law is crucial for understanding the behavior of substances at very low temperatures and has implications for the feasibility of reaching absolute zero. In practical terms, it means that it is impossible to cool a system to absolute zero in a finite number of steps because the entropy of a perfect crystal at absolute zero is exactly zero. This law provides a reference point for the determination of absolute entropies of substances, which are essential for thermodynamic calculations involving chemical reactions and phase changes.

The **Zeroth Law of Thermodynamics**, often considered so fundamental it was added as a "zeroth" law, establishes the concept of temperature and thermal equilibrium. It states that if two systems are each in thermal equilibrium with a third system, then they are in thermal equilibrium with each other. This law forms the basis of temperature measurement and the use of thermometers as it implies that temperature is a transitive property that can be used to compare the thermal state of different systems.

The laws of thermodynamics provide a comprehensive framework that explains not only how energy transformations can be achieved but also the limits within which all physical processes must operate.

Thermodynamic Processes and Energy Changes

In the realm of thermodynamics, understanding the nuances of various processes is crucial for engineers aiming to harness the principles of energy transformation efficiently. These processes, namely isothermal, adiabatic, and isobaric, each describe a unique set of conditions under which a system exchanges heat and work with its surroundings, leading to changes in internal energy, pressure, volume, and temperature.

Isothermal processes occur at a constant temperature, leveraging the fact that the temperature of the system remains unchanged throughout the process. This constancy is achieved by ensuring that the system is in thermal equilibrium with an external reservoir. The first law of thermodynamics, $\Delta U = Q - W$, simplifies to $Q = W$ in an isothermal process for an ideal gas, as there is no change in internal energy ($\Delta U = 0$) due to the constant temperature. The work done by or on the system can be expressed as $W = nRT \ln\left(\frac{V_f}{V_i}\right)$, where n is the amount of substance, R is the ideal gas constant, T is the absolute temperature, and V_f and V_i are the final and initial volumes, respectively.

Adiabatic processes are characterized by the absence of heat exchange between the system and its environment. This implies that all the work done on or by the system results in a change in the internal energy of the system, leading to variations in temperature and volume while the process remains heat-quake ($Q = 0$). The relationship governing adiabatic processes for an ideal gas is given by $PV^\gamma = \text{constant}$, where P is the pressure, V is the volume, and γ is the heat capacity ratio (C_p/C_v), with C_p and C_v being the heat capacities at constant pressure and volume, respectively. The work done in an adiabatic process can be derived from the first law of thermodynamics and is given by $W = \frac{P_f V_f - P_i V_i}{\gamma - 1}$, where P_f and V_f are the final pressure and volume, and P_i and V_i are the initial pressure and volume.

Isobaric processes take place at a constant pressure, allowing the volume and temperature of the system to change while the pressure remains steady. The work done during an isobaric process is directly proportional to the change in volume, $W = P\Delta V$, where P is the constant pressure and ΔV is the change in volume. The heat transferred in such processes can be calculated using the equation $Q = nC_p\Delta T$, where n is the amount of substance, C_p is the heat capacity at constant pressure, and ΔT is the change in temperature.

Each of these thermodynamic processes plays a pivotal role in engineering applications, from the design of heat engines and refrigeration cycles to the analysis of fluid flow and energy systems. By meticulously applying the principles governing these processes, engineers can optimize systems for enhanced performance, energy efficiency, and sustainability, addressing the complex challenges of modern mechanical engineering.

Performance of Thermodynamic Components

Evaluating the performance metrics of thermodynamic components such as turbines, compressors, and nozzles is crucial for understanding their efficiency and operational capabilities in engineering applications. These components are fundamental in various mechanical and aerospace engineering systems, where their performance directly impacts the overall system efficiency and effectiveness.

Turbines convert fluid energy into mechanical work. The performance of turbines can be assessed by parameters such as efficiency, power output, and specific speed. The efficiency of a turbine, often expressed as a percentage, measures how well the turbine converts the energy in the fluid into mechanical work. It is calculated by the ratio of the mechanical power output to the fluid power input. The power output, measured in watts (W) or horsepower (HP), indicates the work produced by the turbine per unit time. Specific speed is a dimensionless parameter that provides a basis for comparing the performance of different turbines under similar conditions.

Compressors are devices used to increase the pressure of a fluid. The key performance metrics for compressors include isentropic efficiency, pressure ratio, and volumetric flow rate. Isentropic efficiency is a measure of the compressor's effectiveness in compressing the fluid without significant energy losses and is defined as the ratio of the work input required for an isentropic compression process to the actual work input. The pressure ratio, which is the ratio of the outlet pressure to the inlet pressure, indicates the compressor's capability to increase fluid pressure. The volumetric flow rate, measured in cubic feet per minute (CFM) or cubic meters per hour (m^3/h), represents the volume of fluid the compressor can process per unit time.

Nozzles are devices designed to control the direction or characteristics of a fluid flow as it exits (or enters) an enclosed chamber or pipe. Performance metrics for nozzles include discharge coefficient, flow rate, and exit velocity. The discharge coefficient is a dimensionless number that compares the actual discharge to the theoretical discharge, indicating the efficiency of the nozzle design. The flow rate, similar to that in compressors, measures the volume of fluid passing through the nozzle per unit time. The exit velocity, measured in meters per second (m/s) or feet per second (ft/s), is the speed of the fluid as it leaves the nozzle, which is critical for applications requiring high-speed jet streams.

Engineers must consider these parameters to optimize the thermodynamic components for specific applications, ensuring that the systems operate at maximum efficiency, meet operational requirements, and adhere to environmental and safety standards.

Power Cycles: Rankine, Brayton, Otto

Power cycles are fundamental to understanding energy conversion systems, especially in the context of power generation. These cycles are the backbone of thermal power plants, internal combustion engines, and refrigeration systems. The Rankine, Brayton, and Otto cycles represent three pivotal concepts in thermodynamics, each illustrating a unique method of converting heat into work or vice versa.

The **Rankine cycle** is predominantly used in steam power plants. It consists of four main processes: isentropic compression in a pump, constant pressure heat addition in a boiler, isentropic expansion in a turbine, and constant pressure heat rejection in a condenser. The efficiency of the Rankine cycle can be enhanced by superheating the steam before expansion and by reheating it between expansion stages. The cycle can be represented by the equation $\eta = 1 - \frac{T_{out}}{T_{in}}$, where η is the efficiency, T_{out} is the temperature of the fluid leaving the system, and T_{in} is the temperature of the fluid entering the system.

The **Brayton cycle**, also known as the Joule cycle, is the foundation of gas turbine engines. It involves adiabatic compression, constant pressure heat addition, adiabatic expansion, and constant pressure heat rejection. The efficiency of the Brayton cycle is improved by increasing the compression ratio and by employing intercooling, reheating, and regeneration. The efficiency equation for the ideal Brayton cycle is given by $\eta = 1 - \left(\frac{P_1}{P_2}\right)^{\frac{\gamma-1}{\gamma}}$, where P_1 and P_2 are the pressures before and after compression, and γ is the specific heat ratio of the gas.

The **Otto cycle** is the ideal cycle for spark-ignition internal combustion engines. It consists of two isentropic processes (adiabatic compression and expansion) and two isochoric processes (constant volume heat addition and rejection). The efficiency of the Otto cycle depends on the compression ratio and the specific heat ratio of the working fluid. The efficiency can be expressed as $\eta = 1 - \left(\frac{V_1}{V_2}\right)^{\gamma-1}$, where V_1 and V_2 are the volumes before and after compression, respectively.

Each of these cycles demonstrates the conversion of heat into work through different processes and under varying conditions. Understanding these cycles allows engineers to design and analyze energy systems for optimal performance, efficiency, and environmental impact. The analysis of these cycles involves a deep understanding of thermodynamic principles, such as the first and second laws of thermodynamics, and the properties of the working fluids involved. By mastering the intricacies of these power cycles, engineers can innovate and improve the design of power generation systems, contributing to the advancement of energy technologies and the reduction of carbon footprints.

Refrigeration and Heat Pump Cycles

Refrigeration and heat pump cycles are pivotal in the domain of thermodynamics, especially when it comes to applications requiring thermal management and energy conversion. These cycles operate on the principle of transferring heat from a cooler to a warmer space, which is contrary to the natural flow of heat, and necessitate external work to achieve this transfer. The coefficient of performance (COP) serves as a primary measure of efficiency for these systems, quantifying the ratio of heating or cooling provided to the work input required.

Refrigeration Cycles typically employ the vapor-compression cycle, which consists of four main processes: evaporation, compression, condensation, and expansion. In the evaporation process, a refrigerant absorbs heat from the refrigerated space, causing it to vaporize. The vaporized refrigerant is then compressed, which increases its temperature and pressure. Next, in the condensation process, the high-pressure vapor releases its heat to the surroundings and condenses into a liquid. Finally, the liquid refrigerant undergoes expansion, reducing its pressure and temperature, and is then returned to the evaporator to repeat the cycle. The COP for refrigeration is defined as $COP_{ref} = \frac{Q_L}{W}$, where Q_L is the heat removed from the refrigerated space, and W is the work input to the compressor.

Heat Pump Cycles share a similar process flow to refrigeration cycles but are utilized for the purpose of heating a space. The key difference lies in the desired output; for heat pumps, the focus is on the heat delivered to the heated space rather than the heat removed from a refrigerated space. Consequently, the COP for heat pumps is given by $COP_{hp} = \frac{Q_H}{W}$, where Q_H is the heat delivered to the heated space. It's noteworthy that the COP of a heat pump is typically higher than that of a refrigeration cycle due to the heat pump's dual role in moving heat out of the refrigerated space and into the heated space.

Both refrigeration and heat pump cycles can be further analyzed through the lens of the second law of thermodynamics, which introduces the concept of entropy and provides a framework for understanding the irreversibilities present in real-world systems. These irreversibilities, such as frictional losses in the compressor and non-ideal gas behavior, reduce the actual COP of the system compared to its ideal, reversible counterpart.

The performance of these cycles is also influenced by the choice of refrigerant, which affects both the environmental impact and the efficiency of the cycle. Recent developments in refrigerant technology have focused on reducing the global warming potential (GWP) and ozone depletion potential (ODP) of these substances, leading to the adoption of more environmentally friendly alternatives.

In engineering applications, optimizing the performance of refrigeration and heat pump cycles involves a careful balance of thermodynamic properties, material selection, and system design. Engineers must consider factors such as the temperature lift (the difference between the evaporating and condensing temperatures), the thermodynamic properties of the refrigerant, and the efficiency of the compressor and other cycle components. By enhancing the efficiency of these cycles, engineers can contribute to energy conservation and reduce the environmental impact of heating and cooling systems.

Nonreacting Gas Mixtures

In the realm of thermodynamics, the study of **nonreacting gas mixtures** is pivotal for understanding the behavior and properties of gases in engineering applications. This section delves into the foundational concepts of nonreacting gas mixtures, focusing on their composition, the concept of partial pressures, and the implications for engineering analysis and design. Nonreacting gas mixtures, as the name suggests, consist of a combination of different gases that do not chemically react with each other under given conditions. The behavior of these mixtures can be predicted and analyzed using the principles of ideal gas laws, extended to accommodate multiple gas components.

The total pressure of a nonreacting gas mixture is the sum of the partial pressures of its components. This principle, known as **Dalton's Law of Partial Pressures**, is mathematically represented as $P_{total} = \sum P_i$, where P_i is the partial pressure of each gas component in the mixture. The partial pressure of each gas is directly proportional to its mole fraction in the mixture, which can be expressed as $P_i = X_i P_{total}$, with X_i being the mole fraction of the ith component. This relationship underscores the significance of mole fraction as a determinant of a gas's contribution to the total pressure of the mixture.

The concept of **partial pressures** is instrumental in applications such as the design of gas separation processes, the analysis of atmospheric gases, and the calculation of gas-phase reactions in combustion. For instance, in air conditioning and refrigeration systems, the performance analysis and design heavily rely on the properties of refrigerant mixtures, where understanding the behavior of nonreacting gas mixtures under different temperatures and pressures is crucial.

Moreover, the **Gibbs-Dalton Law** asserts that the properties of a gas mixture can be determined by summing the contributions of each component gas, treated as if it existed alone at the mixture's total volume and temperature. This simplification is particularly useful in engineering calculations, allowing for the analysis of complex systems by breaking them down into more manageable parts.

In the analysis of **nonreacting gas mixtures**, engineers often employ the **ideal gas equation of state**, extended for mixtures as $P_{total} V = n_{total} RT$, where V is the volume, n_{total} is the total number of moles, R is the universal gas constant, and T is the temperature. This equation facilitates the calculation of the mixture's overall density, specific volume, and other thermodynamic properties essential for the design and analysis of engineering systems.

Understanding the behavior of nonreacting gas mixtures is also fundamental in environmental engineering, where the analysis of pollutant dispersion in the atmosphere requires accurate modeling of gas interactions. Similarly, in the aerospace industry, the performance of propulsion systems depends on the precise calculation of exhaust gas properties, necessitating a thorough understanding of nonreacting gas mixtures.

Mastery of the concepts of partial pressures and the behavior of gas mixtures under various conditions is essential for engineers aiming to innovate and improve the efficiency, safety, and environmental compatibility of engineering systems.

Psychrometrics in HVAC Applications

Psychrometrics is a branch of thermodynamics that deals with the thermophysical properties of moist air and the use of these properties to analyze conditions and processes involving moist air. This field is particularly relevant in HVAC (Heating, Ventilation, and Air Conditioning) applications, where the control of air moisture content is crucial for comfort, process requirements, and equipment operation. The psychrometric chart serves as a valuable tool for engineers, providing a graphical representation of the properties of moist air at various conditions. Understanding the key concepts and properties represented in the psychrometric chart, such as dry bulb temperature, wet bulb temperature, relative humidity, dew point temperature, specific humidity (or mixing ratio), and enthalpy, is essential for the analysis and design of HVAC systems.

Dry Bulb Temperature (DBT) is the temperature of air measured by a thermometer freely exposed to the air but shielded from radiation and moisture. It is the most commonly used measure of air temperature.

Wet Bulb Temperature (WBT) is a measure of the lowest temperature that can be achieved by evaporative cooling of a moistened surface. It is an indicator of the moisture content in the air, with lower wet bulb temperatures corresponding to drier air.

Relative Humidity (RH) is the ratio of the current absolute humidity to the highest possible absolute humidity (which depends on the current air temperature). RH is a critical factor in human comfort, as it affects the efficacy of sweating and cooling.

Dew Point Temperature is the temperature at which air becomes saturated with moisture and dew can form. It is a direct measure of the moisture content in the air and is particularly important in preventing condensation in building structures and HVAC components.

Specific Humidity (or Mixing Ratio) is the weight of water vapor per unit weight of dry air. It is a measure of the actual amount of moisture in the air, regardless of the air's temperature.

Enthalpy of moist air is a measure of its total heat content, including both the sensible and latent heat of the air and water vapor mixture. It is a crucial property in the analysis of energy transfer in HVAC processes.

The psychrometric chart visually represents these properties and their interrelationships, allowing engineers to quickly determine the state of moist air and the changes that occur during heating, cooling, humidification, and dehumidification processes. By plotting the initial and final states of air on the chart, one can easily determine the necessary heating or cooling loads, the amount of moisture to be added or removed, and the expected changes in temperature and humidity. This capability is invaluable in the design and optimization of HVAC systems, ensuring they meet the required comfort and process conditions efficiently.

In HVAC applications, psychrometrics is used to calculate the cooling load, design air conditioning systems, and determine the optimal operation conditions for achieving desired indoor air quality and comfort levels. For instance, in summer, air conditioning systems not only cool the air but also remove excess moisture to maintain comfortable humidity levels. Conversely, in winter, heating systems may need to add moisture to the air to prevent it from becoming too dry. The psychrometric chart facilitates these calculations by providing a comprehensive view of how air properties change with various heating, cooling, humidification, and dehumidification processes.

Moreover, psychrometrics plays a critical role in other applications such as drying processes, where the removal of moisture from materials is required, and in calculating the dew point to prevent condensation on surfaces. In industries where air quality and moisture control are critical, such as pharmaceuticals, food processing, and electronics manufacturing, psychrometric analysis is essential for designing systems that maintain precise environmental conditions.

HVAC Processes and Systems

Heating, Ventilation, and Air-Conditioning (HVAC) systems are integral to maintaining a comfortable and safe indoor environment in buildings. These systems are designed to control the temperature, humidity, and air quality within a space, ensuring optimal conditions for occupants and processes. The design and operation of HVAC systems are governed by the principles of thermodynamics, fluid mechanics, and heat transfer, which together dictate how heat and moisture are moved in and out of the indoor environment.

The primary components of an HVAC system include a heating unit, a cooling unit, and ventilation equipment. The heating unit, often a furnace or a heat pump, provides warmth to the building by burning fuel or converting electricity into heat. The cooling unit, typically an air conditioner or chiller, removes excess heat from the building to lower the indoor temperature. Ventilation equipment, such as fans and ductwork, circulates fresh air within the building and expels stale air to the outside, maintaining air quality and removing excess moisture.

The operation of an HVAC system is based on the principles of thermodynamics, particularly the first and second laws. The first law, the conservation of energy, dictates that energy cannot be created or destroyed, only transferred or converted from one form to another. This principle underlies the operation of HVAC components, as they transfer heat from one place to another, either into the building for heating or out of the building for cooling. The second law of thermodynamics, which states that heat naturally flows from a hotter to a cooler body, is harnessed in HVAC systems to move heat against its natural direction, from inside a building to the outside for cooling, or from the outside air or ground into a building for heating.

The efficiency of HVAC systems is a critical consideration in their design and operation. The Coefficient of Performance (COP) for heating and the Energy Efficiency Ratio (EER) for cooling are key metrics used to evaluate system performance. The COP is the ratio of useful heating provided to the work required to produce that heating, while the EER is the ratio of cooling output divided by electrical energy input. Higher values of COP and EER indicate more efficient systems that provide more heating or cooling output per unit of energy consumed.

Ventilation is another crucial aspect of HVAC systems, ensuring that indoor air quality is maintained to safeguard the health and comfort of building occupants. Ventilation removes contaminants such as dust, carbon dioxide, and volatile organic compounds (VOCs) from indoor air and controls humidity levels to prevent mold growth and maintain structural integrity. The rate of ventilation, typically measured in cubic feet per minute (CFM), must be carefully balanced to provide adequate air quality without unnecessarily increasing energy consumption for heating or cooling the incoming air.

In addition to maintaining comfort and air quality, HVAC systems must also be designed to meet the specific needs of the building and its occupants. This includes considering the local climate, building orientation, occupancy patterns, and thermal properties of the building materials. Advanced HVAC systems may incorporate variable refrigerant flow (VRF) systems, zoning controls, and smart thermostats to optimize energy use and comfort levels dynamically.

The integration of renewable energy sources, such as solar thermal panels or geothermal heat pumps, into HVAC systems is an emerging trend aimed at reducing the environmental impact of heating and cooling buildings. These technologies leverage renewable resources to provide heating or cooling, significantly reducing the reliance on fossil fuels and lowering greenhouse gas emissions.

In conclusion, HVAC systems are complex assemblies that require a deep understanding of thermodynamic principles to design and operate effectively. By carefully selecting and integrating HVAC components, engineers can create systems that not only provide comfort and maintain air quality but also do so in an energy-efficient and environmentally responsible manner.

Combustion Reactions and Products

Combustion is a high-temperature exothermic redox chemical reaction between a fuel and an oxidant, typically atmospheric oxygen, that produces oxidized, often gaseous products, in a mixture termed as smoke. Understanding the combustion process is crucial for mechanical engineers, especially when designing engines and energy systems. The combustion of hydrocarbons, which are the main components of fossil fuels like natural gas, petroleum, and coal, is a primary source of energy in various applications. The complete combustion of a hydrocarbon fuel results in carbon dioxide (CO_2) and water (H_2O), releasing energy. However, incomplete combustion can occur, producing carbon monoxide (CO) and even soot or carbon (C), which are less efficient and more harmful to the environment.

The stoichiometry of the combustion reaction is fundamental in calculating the theoretical air requirement and the composition of the combustion products. For a hydrocarbon fuel represented by the formula C_xH_y, the balanced chemical equation for its complete combustion with oxygen (O_2) is given by:

$$C_xH_y + \left(x + \frac{y}{4}\right) O_2 \rightarrow xCO_2 + \frac{y}{2}H_2O$$

This equation helps in determining the amount of oxygen needed for complete combustion and the volume of the combustion products.

The energy output of the combustion process, known as the heating value or calorific value of the fuel, is a critical parameter for engineering calculations. It is defined as the amount of heat released when a specified amount of fuel is burned completely and the products are returned to a standard state. The higher heating value (HHV) includes the latent heat of vaporization of water in the combustion products, while the lower heating value (LHV) does not. The LHV is more commonly used in engineering calculations, especially for gas turbines and internal combustion engines, as it represents the actual usable energy.

The efficiency of the combustion process is influenced by several factors, including the fuel-to-air ratio, combustion temperature, and pressure. Optimizing these parameters can enhance the energy output and reduce emissions. Engineers use the concept of equivalence ratio (ϕ), defined as the actual fuel-to-air ratio divided by the stoichiometric fuel-to-air ratio, to adjust the combustion process for maximum efficiency. A ϕ value of 1 indicates stoichiometric combustion, while values less than 1 represent lean combustion (excess air) and values greater than 1 indicate rich combustion (excess fuel).

Emissions from combustion, particularly nitrogen oxides (NO_x), sulfur dioxide (SO_2), and particulate matter, are significant environmental concerns. NO_x emissions, for example, are formed through the high-temperature oxidation of nitrogen in the air or the fuel. Controlling these emissions is a major focus of combustion research, leading to the development of technologies like selective catalytic reduction (SCR) and low-NO_x burners.

Chapter 12: Heat Transfer

Conduction: Heat Transfer in Solids

Heat conduction, a fundamental mode of heat transfer, occurs in solid materials when there is a temperature gradient. This process is governed by Fourier's law, which states that the rate of heat transfer through a material is proportional to the negative gradient of the temperature and the area through which the heat is flowing, mathematically expressed as $q = -kA\frac{dT}{dx}$, where q is the heat transfer rate, k is the thermal conductivity of the material, A is the cross-sectional area, and $\frac{dT}{dx}$ is the temperature gradient in the direction of the heat flow. The negative sign indicates that heat flows from higher to lower temperatures.

Thermal conductivity, k, is a material property that quantifies its ability to conduct heat. Materials with high thermal conductivity, such as metals, are excellent heat conductors, while those with low thermal conductivity, such as insulating materials, are poor heat conductors. The value of k is crucial in engineering applications as it affects the design and efficiency of thermal systems.

The heat conduction equation, derived from Fourier's law, enables the calculation of temperature distribution within a solid over time. For steady-state conduction (where temperatures do not change with time), the equation simplifies, and the temperature distribution can be found by solving differential equations subject to boundary conditions. In one-dimensional cases, the equation is $\frac{d^2T}{dx^2} = 0$, leading to a linear temperature gradient in homogeneous materials.

In transient heat conduction, where temperatures vary with time, the heat equation becomes $\frac{\partial T}{\partial t} = \alpha \frac{\partial^2 T}{\partial x^2}$, where α is the thermal diffusivity of the material, a measure of how quickly it responds to changes in temperature. Solving this equation requires initial and boundary conditions and often employs numerical methods for complex geometries and conditions.

The choice of materials, based on their thermal conductivity, and the geometric design of components, are critical factors in ensuring efficient heat transfer and maintaining desired temperature profiles within systems. The analysis of heat conduction is also fundamental in solving multi-dimensional heat transfer problems, where the heat flow is not limited to one direction. In such cases, the heat conduction equation extends to $\nabla^2 T = 0$ for steady-state conditions or $\frac{\partial T}{\partial t} = \alpha \nabla^2 T$ for transient conditions, with ∇^2 representing the Laplacian operator, indicating how the temperature changes in all spatial directions.

Convection Principles and Heat Transfer

Convection is a mode of heat transfer that involves the movement of fluid, which can be either a liquid or a gas, over a surface or through another fluid, resulting in the transfer of heat. This process can be categorized into two main types: natural (or free) convection and forced convection. In natural convection, the movement of the fluid is primarily due to the differences in density caused by temperature gradients within the fluid. Warmer, less dense portions of the fluid rise, while cooler, denser portions sink, creating a natural circulation pattern that facilitates heat transfer. The governing principle behind natural convection can be described by the Boussinesq approximation, which simplifies the Navier-Stokes equations under the assumption that density variations are small and only significant in the gravity term. This leads to the non-dimensional Rayleigh number $Ra = \frac{g \beta (T_s - T_\infty) L^3}{\nu \alpha}$, where g is the acceleration due to gravity, β is the thermal expansion coefficient, T_s is the surface temperature, T_∞ is the fluid temperature far from the surface, L is the characteristic length, ν is the kinematic viscosity, and α is the thermal diffusivity. The Rayleigh number characterizes the regime of natural convection, indicating whether the flow is laminar or turbulent.

Forced convection, on the other hand, involves the movement of fluid induced by external means, such as a pump or fan, across a surface. This type of convection is characterized by the Reynolds number $Re = \frac{UL}{\nu}$, where U is the velocity of the fluid with respect to the object, and L is the characteristic length. The Reynolds number helps in determining the flow regime, whether it is laminar, transitional, or turbulent, which significantly affects the heat transfer coefficient. The Nusselt number $Nu = \frac{hL}{k}$, where h is the convective heat transfer coefficient and k is the thermal conductivity of the fluid, is a dimensionless parameter that relates the convective to the conductive heat transfer across the fluid. For both natural and forced convection, the heat transfer coefficient h is a crucial factor in determining the rate of heat transfer, and it depends on the properties of the fluid, the flow conditions, and the geometry of the surface.

In engineering applications, the choice between natural and forced convection depends on factors such as the required rate of heat transfer, energy efficiency considerations, and the specific constraints of the system design. Natural convection is often utilized in situations where a passive, energy-efficient means of cooling is desired, such as in the design of heat sinks for electronic components. Forced convection is typically employed in situations requiring more effective and controllable heat transfer, such as in automotive radiators, air conditioning systems, and industrial heat exchangers.

The analysis of convection involves solving the fluid flow and heat transfer equations, which can be highly complex, especially in turbulent flow regimes or in geometries with intricate shapes. Computational Fluid Dynamics (CFD) tools are commonly used to simulate convection processes, allowing engineers to predict the performance of heat transfer systems under various operating conditions. These simulations can inform design decisions, such as the selection of materials, the geometry of heat transfer surfaces, and the choice of fluid, to optimize the efficiency and effectiveness of thermal management systems.

Understanding the principles of convection and the factors affecting it is essential for mechanical engineers involved in the design of systems where heat transfer plays a critical role. Mastery of these concepts enables the development of innovative solutions to thermal management challenges, contributing to advancements in a wide range of engineering applications, from energy systems to electronic devices.

Radiation: Heat Transfer by Waves

Radiation heat transfer is a fundamental mode of energy exchange that occurs through electromagnetic waves without the need for a physical medium. This process is pivotal in engineering applications, ranging from thermal management in electronic devices to the design of solar panels and thermal insulation systems. The Stefan-Boltzmann law and the concept of emissivity are central to understanding and quantifying radiation heat transfer. The Stefan-Boltzmann law states that the total energy radiated per unit surface area of a black body is directly proportional to the fourth power of the black body's absolute temperature. Mathematically, it is expressed as $E = \sigma T^4$, where E is the emissive power in watts per square meter (W/m^2), T is the absolute temperature in kelvins (K), and σ is the Stefan-Boltzmann constant, approximately equal to $5.67 \times 10^{-8} W/m^2 K^4$. This law highlights the significant impact of temperature on radiation heat transfer, with the rate of energy emission increasing rapidly with temperature.

Emissivity, denoted as ϵ, is a measure of a material's ability to emit thermal radiation relative to that of a perfect black body. It is a dimensionless quantity ranging between 0 and 1, where 1 represents a perfect black body that emits the maximum possible radiation at a given temperature, and 0 represents a perfect reflector that emits no radiation. The emissive power of a real surface is given by $E = \epsilon \sigma T^4$, indicating that the material's emissivity directly influences its thermal radiation. Materials with high emissivity are efficient radiators and are used in applications requiring effective heat dissipation, while low-emissivity materials are used to minimize heat transfer in thermal insulation.

The relationship between the Stefan-Boltzmann law and emissivity is essential for engineers involved in the design and analysis of systems that utilize thermal radiation. For example, in managing the heat of electronic devices, materials with high emissivity coatings can be employed to improve heat dissipation. On the other hand, in the construction of buildings, materials with low emissivity are often chosen for window glazing to minimize heat loss or gain, thereby enhancing energy efficiency.

The analysis of radiation heat transfer also involves the view factor, which accounts for the geometric relationship between radiating surfaces. The view factor, or configuration factor, quantifies the proportion of radiation leaving one surface that directly reaches another surface, influencing the net radiation exchange between surfaces. Accurate calculation of view factors is essential in complex geometries where multiple surfaces exchange radiation.

Transient Heat Transfer Analysis

Transient processes in heat transfer, particularly unsteady-state conduction, represent a critical area of study for engineers preparing for the FE Mechanical exam. These processes are characterized by temperature changes within a material over time, unlike steady-state conduction where temperatures remain constant. The mathematical foundation for analyzing transient heat conduction is the heat diffusion equation, also known as the transient heat conduction equation. This equation is given by $\frac{\partial T}{\partial t} = \alpha \frac{\partial^2 T}{\partial x^2}$, where T is the temperature, t is time, x is the spatial coordinate, and α is the thermal diffusivity of the material. Thermal diffusivity is a property that combines the thermal conductivity, density, and specific heat capacity of a material, indicating how quickly it responds to changes in temperature.

The solution to the transient heat conduction equation depends on the initial and boundary conditions of the problem. Initial conditions specify the temperature distribution within the material at the start of observation, $t = 0$, while boundary conditions define the temperature and/or heat flux at the material's boundaries. Common types of boundary conditions include specified temperature, specified heat flux, convective, and adiabatic conditions. The complexity of transient heat conduction problems often requires the use of numerical methods for solution, such as the finite difference method, finite element method, or the use of lumped system analysis for simple cases where the Biot number is less than 0.1, indicating that temperature gradients within the object can be neglected.

In engineering practice, understanding transient heat conduction is essential for designing systems that are subject to time-varying thermal conditions. Examples include thermal protection systems for spacecraft during re-entry, cooling of electronic devices, and thermal processing of materials. Engineers must be able to predict how temperatures within a system will change over time to ensure that materials and components can withstand thermal stresses and maintain their integrity and functionality.

The analysis of transient heat transfer also plays a significant role in energy storage systems, where the ability to quickly absorb or release heat is crucial. Phase change materials (PCMs), for example, leverage the latent heat of fusion during phase transitions to store or release large amounts of energy. The effective design of systems incorporating PCMs requires a thorough understanding of transient heat transfer to optimize the charging and discharging processes.

Heat Exchangers Design and Analysis

Heat exchangers are pivotal components in engineering applications, facilitating the transfer of heat between two or more fluids at different temperatures without mixing them. Their design and analysis are critical for optimizing thermal systems in various industries, including power generation, chemical processing, and HVAC systems. The core objective in designing a heat exchanger is to maximize the heat transfer rate while minimizing the pressure drop and material costs. This involves a detailed understanding of the thermal and hydraulic performance of heat exchangers, which can be achieved through the application of fundamental principles of heat transfer and fluid mechanics.

The effectiveness of a heat exchanger is often described by its ability to approach the maximum possible temperature difference between the fluids. This is quantified by the effectiveness ϵ, which is a function of the heat capacity rates of the two fluids and the overall heat transfer coefficient U. The effectiveness is defined as the ratio of the actual heat transfer to the maximum possible heat transfer, given by

$$\epsilon = \frac{Q}{Q_{max}} = \frac{T_{h,in} - T_{h,out}}{T_{h,in} - T_{c,in}}$$

where $T_{h,in}$ and $T_{h,out}$ are the inlet and outlet temperatures of the hot fluid, and $T_{c,in}$ is the inlet temperature of the cold fluid.

The overall heat transfer coefficient U is a measure of the heat exchanger's ability to conduct heat through the materials that separate the two fluids. It is influenced by the thermal conductivities of the heat exchanger materials, the geometry of the heat exchanger, and the convective heat transfer coefficients on both the hot and cold sides. The overall heat transfer coefficient can be calculated from

$$U = \left(\frac{1}{h_h} + \frac{1}{h_c} + \frac{\delta}{k} \right)^{-1}$$

where h_h and h_c are the convective heat transfer coefficients on the hot and cold sides, respectively, δ is the thickness of the wall separating the fluids, and k is the thermal conductivity of the wall material.

The design of a heat exchanger also requires the selection of an appropriate configuration to meet the specific requirements of an application. The most common types of heat exchangers include shell-and-tube, plate-and-frame, and air-cooled. Each type has its unique advantages and limitations in terms of thermal performance, pressure drop, physical size, and cost. For instance, shell-and-tube heat exchangers are widely used in industries due to their robustness and ability to handle high pressures and temperatures. In contrast, plate-and-frame heat exchangers offer higher thermal efficiency and are more compact but may not be suitable for very high-pressure applications.

The analysis of heat exchangers involves determining the temperature distribution and heat transfer rate under various operating conditions. This requires solving the heat exchanger design equation, which relates the heat transfer rate to the overall heat transfer coefficient, the heat transfer area, and the log mean temperature difference (LMTD) between the hot and cold fluids. The LMTD is calculated from

$$\Delta T_{lm} = \frac{\Delta T_1 - \Delta T_2}{\ln \left(\frac{\Delta T_1}{\Delta T_2} \right)}$$

where ΔT_1 and ΔT_2 are the temperature differences between the hot and cold fluids at each end of the heat exchanger.

In conclusion, the design and analysis of heat exchangers are complex but essential tasks that require a deep understanding of heat transfer principles, fluid dynamics, and material science. By carefully selecting the type, configuration, and materials of a heat exchanger, engineers can significantly improve the efficiency and reliability of thermal systems, leading to energy savings and reduced operational costs.

Chapter 13: Measurements and Controls

Sensors and Transducers

Sensors and transducers play a pivotal role in the engineering world, converting physical quantities into electrical signals for precise measurement and control. These devices are the critical interface between the physical world and the digital systems that monitor and manage engineering processes. Understanding their operation, characteristics, and applications is essential for any engineer preparing for the FE Mechanical exam.

Sensors are devices that detect changes in physical quantities such as temperature, pressure, displacement, and velocity, and then respond with an output signal. Transducers, on the other hand, convert one form of energy into another, often transforming a signal associated with a physical quantity into an electrical signal. This distinction is crucial because it underlines the broad range of devices and principles that an engineer must be familiar with.

The operation of sensors and transducers involves several key principles. For instance, a thermocouple, which measures temperature, operates based on the Seebeck effect, where a voltage is generated across two different metals that are joined at one end and exposed to a temperature gradient. The generated voltage is then proportional to the temperature difference. This principle is represented by the formula $V = S \cdot \Delta T$, where V is the voltage generated, S is the Seebeck coefficient, and ΔT is the temperature difference between the two junctions.

Pressure sensors, another critical type of sensor, can operate based on piezoelectric effects, where mechanical stress upon certain materials generates an electrical charge. The relationship between the applied pressure P and the generated charge Q can be described by $Q = d \cdot P$, where d is the piezoelectric coefficient. This principle is fundamental in designing devices that measure fluid dynamics and static pressures within mechanical systems.

In the realm of displacement and position measurement, Linear Variable Differential Transformers (LVDTs) are widely used. These devices operate on the principle of mutual inductance change in response to the movement of a magnetic core. The output voltage V_{out} from an LVDT is a function of the core's position x, and can be described by the linear approximation $V_{out} = k \cdot x$ for small displacements, where k is a constant that depends on the transformer's geometry and the excitation voltage.

For velocity and acceleration, piezoelectric accelerometers are commonly employed. These devices generate a charge in response to acceleration, which can then be converted into a voltage signal proportional to the rate of velocity change. The relationship between acceleration a, mass m, and the generated charge Q is given by $Q = m \cdot a$, highlighting the direct correlation between physical movement and electrical signal generation. From the basic thermocouples to sophisticated piezoelectric sensors, the ability to accurately convert physical quantities into electrical signals is foundational in modern engineering practices, ensuring precision, efficiency, and reliability in various mechanical systems.

Control Systems Basics

Control systems are fundamental in mechanical engineering, enabling the automation and regulation of various processes and machines. At the heart of these systems lies the concept of **feedback**, a mechanism by which the system can adjust its operation based on the output to maintain the desired performance. Feedback can be either positive or negative, with negative feedback being more common in engineering applications due to its ability to stabilize systems and reduce errors. The basic structure of a control system can be visualized through **block diagrams**, which represent the system's components and their interactions. These diagrams are crucial for understanding how signals flow through the system and for identifying potential points of failure or inefficiency.

The design and analysis of control systems require a deep understanding of several key principles, including **transfer functions**, **system stability**, and **frequency response**. The transfer function, represented as $G(s) = \dfrac{Y(s)}{X(s)}$, where $Y(s)$ is the output and $X(s)$ is the input, both in the Laplace domain, encapsulates the system's behavior. It allows engineers to analyze how the system responds to different inputs without solving the differential equations governing the system's dynamics directly.

System stability is another critical aspect, often assessed using methods such as the Routh-Hurwitz criterion or the Nyquist stability criterion. These methods enable engineers to determine whether a system will reach a steady state or oscillate indefinitely in response to a disturbance. A stable system is one where the output returns to equilibrium after a disturbance, a fundamental requirement for most engineering applications.

Frequency response analysis, including Bode plots and Nyquist plots, provides insights into how a system reacts to inputs of varying frequencies. This analysis is essential for designing systems that can handle a wide range of operating conditions without exhibiting undesirable behavior such as excessive vibration or noise.

In the context of mechanical engineering, control systems find applications in numerous areas, from regulating the temperature and pressure in HVAC systems to controlling the speed and torque of electric motors. The principles of feedback, system stability, and frequency response are applied to ensure these systems operate efficiently, safely, and reliably.

To design an effective control system, engineers must first define the desired performance characteristics, such as response time, accuracy, and stability margins. They then select appropriate sensors and actuators, design the control algorithm, and simulate the system's performance under various conditions. This iterative process involves adjusting the control strategy and parameters until the system meets the specified requirements.

Control systems are crucial in mechanical engineering, facilitating the accurate management of machinery and processes. The principles of feedback, block diagrams, transfer functions, system stability, and frequency response are vital for the design and analysis of these systems. Through the application of these principles, engineers can develop control systems that improve the efficiency, safety, and reliability of mechanical systems.

Dynamic System Response

Dynamic system response is a critical aspect of control systems, focusing on how systems react over time to various inputs. This concept is pivotal in mechanical engineering, as it directly influences the design and functionality of numerous mechanical and electromechanical systems. The analysis of dynamic system response involves understanding the time-dependent behavior of systems under the influence of external stimuli, which can be deterministic or random in nature. The primary goal is to predict the system's response to inputs, ensuring stability and desired performance under all operating conditions.

The mathematical foundation for analyzing dynamic system response is grounded in differential equations, which describe the relationship between a system's input, its inherent characteristics, and its output over time. The solution to these equations provides insight into the system's transient and steady-state behavior. Transient response refers to the system's reaction immediately following a change in input, characterized by parameters such as rise time, overshoot, settling time, and decay rate. Steady-state response, on the other hand, describes the system's output after it has stabilized from an initial transient phase.

To systematically analyze dynamic system response, engineers employ the Laplace transform, a powerful tool that converts differential equations into algebraic equations in the s-domain. This transformation simplifies the process of solving for system response, making it easier to analyze complex systems. The transfer function, $G(s) = \dfrac{Y(s)}{X(s)}$, represents the system's output to input ratio in the Laplace domain, encapsulating both the transient and steady-state characteristics of the system.

Stability is a paramount concern in dynamic system response analysis. A system is considered stable if its output returns to equilibrium or a steady state after a disturbance. The stability of a system can be assessed using various criteria, such as the Routh-Hurwitz criterion, which examines the roots of the system's characteristic equation, or the Nyquist criterion, which analyzes the system's frequency response. These methods help identify conditions under which a system may become unstable, guiding the design of control strategies to mitigate such risks.

Frequency response analysis is another critical aspect of understanding dynamic system response. It involves studying how a system reacts to sinusoidal inputs of varying frequencies. Tools like Bode plots and Nyquist plots visualize a system's gain and phase shift across a range of frequencies, providing insights into resonance phenomena, bandwidth, and the system's ability to filter or amplify certain frequency components.

In practical applications, engineers must consider the dynamic response of systems in the design of feedback control loops, vibration suppression mechanisms, and in predicting the behavior of mechanical structures under dynamic loads. For instance, in designing an automotive suspension system, engineers must ensure that the system can absorb road irregularities efficiently, minimizing transient vibrations and ensuring passenger comfort while maintaining vehicle stability and control.

Measurement Uncertainty

Measurement uncertainty is a critical concept in engineering that quantifies the doubt associated with a measurement. Understanding and managing measurement uncertainty is essential for engineers to make informed decisions based on measurement results. The sources of measurement uncertainty can be broadly classified into two categories: systematic errors and random errors. Systematic errors, also known as biases, are errors that consistently occur in the same direction whenever a measurement is made. These errors can often be identified and corrected by calibrating measurement instruments against known standards. Random errors, on the other hand, are caused by unpredictable fluctuations in the measurement process and are characterized by their random distribution around the true value. The combined effect of these errors determines the total uncertainty of a measurement.

To quantify measurement uncertainty, engineers use statistical methods. The standard deviation of a series of measurements is a common measure of the variability due to random errors. However, to fully characterize the measurement uncertainty, one must also consider the uncertainty contributions from systematic errors. This comprehensive approach to uncertainty analysis involves identifying all potential sources of uncertainty, quantifying each source, and combining them using the root sum square method. The formula for combining multiple uncertainty components U is given by

$$U = \sqrt{\sum u_i^2}$$

, where u_i represents the individual uncertainty components. This aggregated uncertainty provides a more accurate representation of the confidence that can be placed in a measurement result.

Accuracy and precision are two fundamental aspects of measurement that are often confused but have distinct meanings. Accuracy refers to the closeness of a measured value to the true value, while precision refers to the repeatability of measurements, or how close the measurements are to each other, regardless of their closeness to the true value. A measurement system can be precise without being accurate if it consistently yields results that are far from the true value. Conversely, a measurement can be accurate but not precise if the measurements are close to the true value but scattered widely. The goal in measurement is to achieve both high accuracy and high precision, minimizing both systematic and random errors.

Significant figures play a crucial role in conveying the precision of a measurement. The number of significant figures in a reported measurement reflects the confidence in the measurement's precision. It is important to follow the rules of significant figures when performing calculations to ensure that the precision of the final result is not overstated. When combining measurements with different degrees of precision, the rule of thumb is that the final result should not have more significant figures than the least precise measurement.

Chapter 14: Mechanical Design Analysis

Stress Analysis of Machine Elements

Stress analysis of machine elements is a fundamental aspect of mechanical design that involves evaluating the stress distribution within components subjected to external loads or environmental conditions. This analysis is crucial for ensuring the durability, reliability, and safety of machine parts in their intended applications. The primary goal is to identify stress concentrations, predict failure points, and optimize the design to withstand operational demands. The process incorporates various analytical, numerical, and experimental methods to assess stress levels and their implications on the performance and lifespan of machine elements.

Analytical methods often start with the application of basic principles of mechanics of materials. For instance, when dealing with simple geometries and load conditions, formulas derived from the theory of elasticity can be directly applied. The stress σ in a loaded member can be calculated using $\sigma = \dfrac{F}{A}$, where F is the applied force and A is the cross-sectional area. For more complex situations involving bending, torsion, or combined loading, the stress analysis might involve the use of Bernoulli's beam theory, the torsion formula $\tau = \dfrac{T \cdot r}{J}$, where τ is the shear stress, T is the applied torque, r is the radial distance from the center, and J is the polar moment of inertia, or other relevant equations that consider the geometry and material properties of the component.

Numerical methods, particularly Finite Element Analysis (FEA), have become indispensable in stress analysis due to their ability to handle complex geometries and loading conditions that are challenging to solve analytically. FEA divides the component into a mesh of smaller, simpler elements where the stress-strain relationships can be solved using matrix methods. This approach provides detailed insights into the stress distribution and identifies regions of high stress concentration that could be potential failure points. The accuracy of FEA results significantly depends on the quality of the mesh, the material model used, and the boundary conditions applied, making it essential to validate these models with experimental data or analytical solutions when possible.

Experimental methods, including strain gauging and photoelasticity, offer direct measurements of stress and strain in physical components under load. Strain gauges, when attached to the surface of a component, provide localized strain measurements that can be correlated to stress using Hooke's law for linear elastic materials. Photoelasticity uses models made from optically sensitive materials to visualize stress distribution patterns through the observation of birefringence patterns under polarized light. These experimental techniques are particularly useful for validating analytical and numerical models and for conducting failure analysis.

Material properties play a critical role in stress analysis, as the strength, ductility, and fatigue resistance of the material determine its ability to withstand applied stresses without failure. The selection of materials based on their mechanical properties, such as yield strength, ultimate tensile strength, and modulus of elasticity, is a critical step in the design process. Additionally, understanding the material's behavior under different loading conditions, such as cyclic loading in fatigue analysis, is essential for predicting the component's lifespan and ensuring its safe operation.

Stress analysis of machine elements is a multi-faceted process that integrates principles from mechanics of materials, numerical simulation techniques, and experimental validation to assess the structural integrity of components under load. By accurately determining stress distributions and identifying potential failure points, engineers can design machine elements that meet the required performance criteria while ensuring safety and reliability throughout the component's operational life. This comprehensive approach to stress analysis is vital for advancing mechanical design and achieving the objectives of durability, efficiency, and safety in engineering applications.

Failure Theories and Analysis

In the realm of mechanical engineering, understanding failure theories and analysis is paramount for predicting and preventing failure in materials and components. This section delves into the fundamental theories that underpin failure analysis, providing a comprehensive overview of how these theories are applied to assess the risk of failure under various loading conditions. The primary failure theories discussed include the Maximum Normal Stress Theory, the Maximum Shear Stress Theory, the Distortion Energy Theory, and the Mohr's Criterion for brittle materials.

The Maximum Normal Stress Theory, also known as Rankine's Theory, posits that failure occurs when the maximum normal stress in a material reaches the material's ultimate tensile strength (UTS) or ultimate compressive strength (UCS), depending on the nature of the stress (tensile or compressive). This theory is particularly applicable to brittle materials that fail without significant deformation. The criterion for failure under this theory can be mathematically expressed as $\sigma_{max} = \sigma_{UTS}$ for tensile stresses or $\sigma_{max} = \sigma_{UCS}$ for compressive stresses.

The Maximum Shear Stress Theory, known as Tresca's Criterion, suggests that failure occurs when the maximum shear stress in the material exceeds the shear strength of the material, which is typically half the yield strength ($\tau_{max} = \dfrac{\sigma_y}{2}$). This theory is most applicable to ductile materials that fail after yielding. It is particularly useful for predicting yielding under complex loading conditions.

The Distortion Energy Theory, or von Mises Criterion, is based on the assumption that failure occurs when the distortion energy per unit volume due to applied stresses reaches the distortion energy per unit volume at yield in a uniaxial tensile test. This theory provides a more accurate prediction of failure for ductile materials under multiaxial stress states and is expressed mathematically as $\sqrt{\dfrac{3}{2}}(\sigma_{ij} - \sigma_{mean}) = \sigma_y$, where σ_{ij} represents the stress components and σ_{mean} is the mean normal stress.

Mohr's Criterion for brittle materials takes into account the effect of normal stresses on failure due to shear. It posits that failure occurs when the stress state at a point reaches the failure envelope in the Mohr's stress circle representation. This criterion is particularly effective for analyzing failure in brittle materials under complex stress states, including the presence of tensile stresses which significantly affect their strength.

In applying these theories to predict failure in engineering components, engineers must first accurately determine the stress state within the material or component under the expected loading conditions. This often involves complex calculations or finite element analysis (FEA) to model the stress distribution. Once the stress state is known, the appropriate failure theory can be applied to assess whether the stress conditions exceed the material's failure criteria.

It is crucial to select the most appropriate failure theory based on the material type (ductile or brittle) and the nature of the loading (uniaxial, biaxial, or triaxial). For instance, the Maximum Normal Stress Theory may be more suitable for brittle materials like ceramics and glass, while the Distortion Energy Theory may be better suited for ductile metals subjected to complex loading conditions.

Moreover, understanding the limitations and assumptions inherent in each failure theory is essential for accurate failure prediction. For example, the Maximum Shear Stress Theory does not account for the effect of hydrostatic pressure, while the Distortion Energy Theory assumes isotropic and homogeneous material properties.

Failure theories and analysis play a critical role in the design and assessment of mechanical components, enabling engineers to predict potential failure points and mitigate risks through design optimization and material selection. By applying these theories judanalyzerically and considering the specific conditions and limitations of each, engineers can enhance the reliability and safety of mechanical systems.

Deformation and Stiffness

In the realm of mechanical design and analysis, understanding the concepts of **deformation** and **stiffness** is paramount for ensuring the structural integrity and performance of materials under various loads. Deformation refers to the change in shape or size of a material when subjected to external forces, characterized by strain, which is the ratio of the change in dimension to the original dimension. Stiffness, on the other hand, is a material's resistance to deformation, quantitatively defined by the modulus of elasticity, E, which represents the ratio of stress (force per unit area) to strain in the linear elastic region of the material's stress-strain curve.

The relationship between stress σ and strain ϵ in this linear region is given by Hooke's Law, $\sigma = E\epsilon$, highlighting how stiffness, or the modulus of elasticity, directly influences a material's ability to withstand applied forces without significant deformation. This fundamental principle is critical in selecting materials for engineering applications, where the balance between flexibility and rigidity must be carefully managed to meet design specifications and functional requirements.

Moreover, the analysis of deformation under various loading conditions—tensile, compressive, shear, and torsional—requires a comprehensive understanding of stress-strain relationships and the material properties that influence them. For instance, the shear modulus G and bulk modulus K are also critical parameters in assessing material behavior under shear and volumetric stress, respectively, further complicating the selection process for engineering materials.

In practical applications, engineers must also consider factors such as creep, which is the time-dependent slow deformation under a constant load, and fatigue, which involves material weakening due to repetitive loading cycles. These phenomena can significantly affect the long-term performance and reliability of materials in structural applications, making the understanding of deformation and stiffness not just a matter of theoretical interest but a critical component of engineering design and analysis.

The interplay between deformation and stiffness also extends to the analysis of composite materials, where the arrangement of different materials can be engineered to achieve desired properties that are not possible with individual components alone. This includes the strategic use of high-stiffness fibers in a softer matrix to create composite materials with an optimal balance of strength, stiffness, and weight for specific applications.

Springs: Design and Analysis

Springs are fundamental components in mechanical systems, serving to absorb energy, maintain force, or provide resilience between contacting surfaces. The design and analysis of springs require a thorough understanding of their load capacity and deformation characteristics to ensure they meet the specific requirements of an application. The load capacity of a spring is defined by its ability to withstand forces without permanent deformation or failure, while deformation characteristics refer to the spring's response to applied loads, typically described by its stiffness or spring constant k, and its deflection x.

The spring constant k is a measure of the spring's stiffness, calculated as the ratio of the force F applied to the spring to the resulting displacement x, expressed by the formula $k = \frac{F}{x}$. This relationship is central to Hooke's Law, which states that the force required to extend or compress a spring by some distance x scales linearly with that distance, as long as the spring's elastic limit is not exceeded. The elastic limit is the maximum stress or force per unit area a spring can withstand without undergoing permanent deformation.

In the context of mechanical design, engineers must also consider the material properties of the spring, such as the modulus of elasticity E, shear modulus G, and the ultimate tensile strength, to predict how the spring will behave under various loading conditions. These material properties, along with the spring's geometry, influence its load capacity and deformation characteristics. For example, the Wahl factor, an important consideration in the design of helical springs, accounts for the curvature and direct shear effects, providing a more accurate calculation of the spring constant for closely coiled springs.

The analysis of springs extends beyond simple linear models to include complex behaviors such as buckling in compression springs, resonance frequencies in vibrational applications, and fatigue life under cyclic loading. Engineers use these analyses to design springs that can reliably meet operational requirements, considering factors such as expected life cycles, environmental conditions, and potential failure modes.

For torsion springs, the torque T applied to the spring is proportional to the angle of twist θ, described by the equation $T = k\theta$, where k is the torsional spring constant. This relationship allows for the design of springs that can store rotational energy or apply a rotational force at a known rate.

In applications where springs are subjected to variable or cyclic loads, the S-N curve, representing the relationship between the stress amplitude and the number of cycles to failure, becomes a critical design consideration. This curve helps engineers predict the fatigue life of the spring, ensuring that the spring can withstand the expected number of load cycles over its operational life.

Selecting the appropriate spring for a given application involves balancing these load capacity and deformation characteristics with the constraints of the mechanical system, including space limitations, weight considerations, and cost. By understanding the fundamental principles of spring design and analysis, engineers can develop solutions that optimize performance, durability, and reliability in mechanical systems.

Pressure Vessels and Piping Safety

Pressure vessels and piping systems are critical components in various industrial applications, including the petrochemical, nuclear, and food processing industries. These systems are designed to operate under high pressure and temperature conditions, making the understanding of stress analysis and safety considerations paramount for engineers. The design and analysis of pressure vessels and piping systems must adhere to stringent standards and codes, such as the ASME Boiler and Pressure Vessel Code (BPVC) and the B31 Piping Codes, to ensure operational safety and integrity.

The stress analysis of pressure vessels involves the calculation of stresses under internal pressure, external pressure, and other loading conditions such as weight and thermal gradients. The primary stresses of concern are circumferential (hoop) stress σ_h and longitudinal stress σ_l. The hoop stress, which is the stress in the circumferential direction, is given by the formula $\sigma_h = \dfrac{pD}{2t}$, where p is the internal pressure, D is the inner diameter of the vessel, and t is the wall thickness. The longitudinal stress, on the other hand, is half of the hoop stress, $\sigma_l = \dfrac{pD}{4t}$, due to the equilibrium of forces in the axial direction. These stresses must not exceed the allowable stress values for the material, which are determined based on the material's yield strength and a safety factor.

Safety considerations in the design of pressure vessels and piping systems extend beyond stress analysis to include the prevention of failure modes such as brittle fracture, fatigue, and corrosion. Brittle fracture is of particular concern in vessels operating at low temperatures or subject to dynamic loading, where the material may fail suddenly without significant deformation. To mitigate this risk, materials with adequate toughness are selected, and impact testing is performed as part of the material specification process.

Fatigue failure, resulting from cyclic loading, is addressed by designing for a sufficient fatigue life based on the expected number of cycles and the stress range. The use of fatigue curves, which relate stress amplitude to the number of cycles to failure, allows engineers to predict the fatigue life of components under varying stress conditions.

Corrosion, a common degradation mechanism in pressure vessels and piping systems, can significantly reduce the wall thickness and, consequently, the pressure-carrying capacity of the system. Corrosion allowances are incorporated into the design to account for expected material loss over the service life of the vessel or piping system. Additionally, material selection and the application of protective coatings are critical strategies for corrosion control.

The integrity of welded joints, which are common in the construction of pressure vessels and piping systems, is another critical safety consideration. Non-destructive testing (NDT) methods such as radiography, ultrasonic testing, and magnetic particle inspection are employed to detect defects in welds that could compromise the structural integrity of the system.

Bearings: Types and Design Principles

Bearings play a pivotal role in mechanical systems by facilitating smooth motion and supporting loads while minimizing friction. Understanding the types of bearings and their design principles is essential for engineers to select the appropriate bearing for a given application, ensuring operational efficiency and longevity of mechanical components. Bearings can be broadly classified into two main categories: rolling element bearings and plain bearings. Each type has its unique advantages and applications, dictated by factors such as load capacity, speed, and environmental conditions.

Rolling element bearings, also known as antifriction bearings, utilize balls or rollers to minimize surface contact and friction between moving parts. These bearings are further categorized into ball bearings, which use spherical balls as the rolling element, and roller bearings, which employ cylindrical or tapered rollers. The choice between ball and roller bearings depends on the specific requirements of the application, including load direction and magnitude, operational speed, and precision. Ball bearings are typically used in applications requiring low friction and high-speed operation, while roller bearings are suited for applications with high radial or axial loads.

Plain bearings, also referred to as sleeve bearings or bushings, rely on sliding motion between surfaces to support loads and allow movement. These bearings are often simpler in design and construction compared to rolling element bearings, making them cost-effective for many applications. Plain bearings can be constructed from various materials, including bronze, plastic, and composite materials, each offering different properties in terms of friction, wear resistance, and load-carrying capacity. The selection of material is critical and should be based on the operating conditions, such as temperature, lubrication availability, and potential exposure to corrosive elements.

The design principles of bearings focus on optimizing performance and longevity while minimizing maintenance requirements. **Load capacity** is a fundamental consideration, requiring engineers to accurately calculate the expected loads the bearing will encounter during operation. Bearings must be selected based on their dynamic load rating, which considers the bearing's ability to withstand load fluctuations and rotational speeds over a specified life expectancy. **Lubrication** is another critical design aspect, essential for reducing friction and wear between moving parts. The choice of lubrication, whether it be grease, oil, or solid lubricants, depends on the bearing type, application speed, and environmental conditions. Proper lubrication reduces the risk of premature failure and extends the bearing's service life.

Sealing is also an important design consideration, especially for bearings operating in environments prone to contamination by dust, moisture, or other foreign materials. Seals protect the bearing from contaminants while retaining lubricant, crucial for maintaining optimal performance and reliability. **Mounting and alignment** are critical for ensuring that bearings operate efficiently. Incorrect mounting and misalignment can lead to uneven load distribution, increased friction, and accelerated wear, ultimately resulting in bearing failure.

Power Screws: Load and Torque Requirements

Power screws are a fundamental element in mechanical design, serving as devices to convert rotational motion into linear motion and vice versa. They are widely used in various applications such as vises, presses, and lifting mechanisms. The design and analysis of power screws focus on their load-carrying capacity and the torque requirements to ensure efficiency and reliability in their operation. The load-carrying capacity of a power screw is determined by its ability to support axial loads without experiencing excessive stress or deformation. This capacity is influenced by the screw's material properties, geometry, and the presence of threads. The axial load F_a that a power screw can support is given by the equation $F_a = \sigma_a \cdot A$, where σ_a is the allowable stress for the screw material and A is the stress area of the screw thread.

The torque required to raise or lower a load using a power screw, T, is a critical parameter for the design of screw-driven systems. This torque depends on several factors including the screw's geometry, the coefficient of friction between the threads, the type of threads (square, trapezoidal, etc.), and the presence of lubrication. The basic equation to calculate the torque required to raise a load is given by $T = (F_a \cdot d_m/2) \cdot (\tan(\lambda) + \mu \cdot \sec(\alpha))$, where d_m is the mean diameter of the screw, λ is the lead angle, μ is the coefficient of friction, and α is the half-angle of the thread. Conversely, the torque required to lower a load includes the frictional forces acting against the motion and is calculated by modifying the friction term in the equation accordingly.

Efficiency of a power screw is another important consideration, defined as the ratio of the work done in raising the load to the work done on the screw. It is a measure of how effectively the input energy is converted into useful work and is affected by the friction between the screw threads and the nut. The efficiency η can be expressed as $\eta = \dfrac{\tan(\lambda)}{\tan(\lambda) + \mu \cdot \sec(\alpha)}$, highlighting the inverse relationship between efficiency and friction. High efficiency is desirable for reducing the power losses and the heat generated during operation, which can be achieved by selecting materials with lower coefficients of friction, applying lubrication, and optimizing the thread geometry.

The selection of a power screw for a specific application requires a careful balance between the load-carrying capacity, the required torque, and the efficiency. Engineers must consider the operational conditions such as the magnitude and direction of the load, the desired speed of movement, the available power source, and environmental factors that may affect performance, such as temperature and contamination. Additionally, the wear resistance of the screw material and the potential for backlash, which is the lost motion or looseness between the threads of the screw and the nut, must be addressed to ensure long-term reliability and precision of the screw-driven system.

In mechanical design analysis, the application of power screws necessitates a comprehensive understanding of their mechanical properties and operational characteristics. By accurately calculating the load-carrying capacity, required torque, and efficiency, and by selecting appropriate materials and thread designs, engineers can effectively utilize power screws in a wide range of mechanical applications, optimizing performance and durability.

Power Transmission Elements

Power transmission in mechanical systems is a critical aspect of mechanical design and analysis, focusing on the efficient transfer of energy from one part of a system to another to perform work. The elements involved in power transmission include gears, belts, and chains, each serving a unique function and application within various mechanical systems. **Gears** are rotating machine elements that transmit torque and motion through interlocking teeth. The interaction between gears allows for significant flexibility in the adjustment of speed, torque, and direction of movement. The fundamental equation governing gear operation is the gear ratio, $i = \frac{N_2}{N_1}$, where N_1 and N_2 are the speeds of the driving and driven gears, respectively. This ratio is inversely proportional to the torque ratio, demonstrating how gears can be utilized to increase torque while decreasing speed, or vice versa. The design considerations for gears include the type (spur, helical, bevel, or worm), material, and the precision of manufacturing to ensure smooth operation and longevity.

Belts and **pulleys** offer another method for transmitting power between shafts that may not be aligned or are at a distance from each other. Belts, typically made from flexible materials like rubber or synthetics, run over pulleys to transfer power through frictional forces between the belt and the pulley surfaces. The tension in the belt must be adequately maintained to prevent slippage, which can be achieved through tensioning mechanisms or by using self-tensioning designs. The velocity ratio of the belt-driven system is determined by the diameters of the driving and driven pulleys, $v = \frac{D_1}{D_2}$, where D_1 and D_2 are the diameters of the driving and driven pulleys, respectively. Belt systems are favored for their quiet operation, low maintenance, and the ability to absorb shock loads.

Chains and **sprockets** provide a means of power transmission through a series of linked segments or rollers, engaging with the teeth of a sprocket to convert rotational motion into linear motion, or vice versa. Unlike belts, chain drives do not slip under normal conditions, offering a consistent velocity ratio that is crucial for applications requiring precise timing, such as in synchronous engines and conveyance systems. The design of chain drives considers the pitch of the chain, the number of teeth on the sprocket, and the center distance between the sprockets to ensure efficient operation and to minimize wear.

Each of these power transmission elements has its advantages and limitations, influenced by factors such as the required speed and torque, the presence of misalignment, environmental conditions, and maintenance requirements. The selection of the appropriate power transmission element is critical to the design of efficient, reliable, and cost-effective mechanical systems. Engineers must carefully consider the operational requirements, including load characteristics, speed variations, and spatial constraints, to determine the most suitable power transmission method. Additionally, the analysis of power transmission elements involves understanding the principles of kinematics and dynamics, material science, and mechanical design to predict performance, optimize design, and ensure the longevity and reliability of mechanical systems.

Joining Methods

In the realm of mechanical design and analysis, understanding the various joining methods is crucial for the assembly of components in engineering applications. These methods, including welding, adhesives, and fasteners, are selected based on factors such as the materials being joined, the required strength of the joint, cost considerations, and the intended application environment.

Welding is a process that joins materials, usually metals or thermoplastics, by causing coalescence. This is often done by melting the workpieces and adding a filler material to form a pool of molten material that cools to become a strong joint. There are several types of welding, including arc welding, gas welding, and resistance welding, each with its specific applications and advantages. For instance, arc welding is suitable for joining heavy steel sections, whereas gas welding is often used for repair work due to its portability.

Adhesives offer a versatile joining solution that can bond a wide range of materials, including metals, plastics, and composites. The use of adhesives allows for the distribution of stress over a broad area, reducing stress concentration at any single point. This method is particularly beneficial in applications where aesthetics are important, as it does not require the surface alteration of the workpiece. However, the strength of adhesive joints can be affected by temperature and moisture, and proper surface preparation is critical for achieving a durable bond.

Fasteners, such as bolts, screws, and rivets, provide a non-permanent means of joining materials that can be easily assembled and disassembled. This method is advantageous for structures that require maintenance or modifications, as it allows for easy access to the components. The selection of fasteners depends on several factors, including the types of materials being joined, the forces acting on the joint, and environmental conditions. For example, stainless steel fasteners are preferred for outdoor applications due to their corrosion resistance.

Each joining method has its specific considerations, such as weldability of materials in welding, surface preparation in adhesive bonding, and the mechanical properties of fasteners. Engineers must evaluate these factors in the context of their design requirements to select the most appropriate joining method for their application. This decision-making process involves a thorough understanding of the mechanics of materials, as well as the principles of mechanical design and analysis, to ensure the reliability and performance of the assembled structure.

Design for Manufacturability

Design for Manufacturability (DFM) is a critical engineering practice aimed at reducing production costs, enhancing product quality, and ensuring ease of manufacturing. This approach necessitates a comprehensive understanding of manufacturing processes, material properties, and the capabilities and limitations of manufacturing equipment. By integrating DFM principles early in the design phase, engineers can identify and mitigate potential manufacturing issues, streamline assembly processes, and reduce the need for costly redesigns.

Tolerances are a fundamental aspect of DFM, defining the acceptable limits of variation for the dimensions and physical properties of components. Establishing appropriate tolerances is a balancing act; overly tight tolerances may increase manufacturing complexity and costs without significantly improving product performance, while too loose tolerances can lead to assembly issues or functional deficiencies. Engineers must consider the specific manufacturing processes and the inherent variability of each method when setting tolerances. For instance, injection molding will have different tolerances compared to CNC machining due to differences in process capabilities and material behavior.

Limits refer to the maximum and minimum dimensions that parts can have to ensure they function correctly and fit together as intended. Limits are closely tied to tolerances, with the difference between the upper and lower limits defining the tolerance range. In setting limits, engineers must account for the cumulative effect of tolerances on assemblies, known as tolerance stack-up, which can significantly impact manufacturability and product performance.

Fit standards are guidelines that dictate how parts will fit together and operate in an assembly. Fits can range from loose to tight, with different standards for different types of fits, such as clearance fits, interference fits, and transition fits. The selection of fit standards is crucial for ensuring that assemblies can be easily put together and will function as intended under all operating conditions. For example, a shaft and bearing may require a precise interference fit to prevent relative motion under load, while a clearance fit might be appropriate for components that need to move freely relative to one another.

Incorporating DFM principles requires a multidisciplinary approach, involving collaboration between design engineers, manufacturing engineers, and quality assurance teams. Simulation tools and prototyping can be invaluable in identifying manufacturability issues early in the design process. For example, finite element analysis (FEA) can predict how variations in material properties and manufacturing processes might affect component performance, while rapid prototyping techniques, such as 3D printing, can be used to quickly evaluate design concepts and test fits and tolerances.

Ultimately, successful implementation of DFM principles can lead to significant cost savings, improved product quality, and shorter time to market. By considering manufacturing constraints and opportunities from the outset, engineers can design products that are not only innovative and functional but also economical to produce and assemble.

Quality and Reliability Principles

In the realm of mechanical design and analysis, the concepts of **quality** and **reliability** are paramount for ensuring the long-term performance of components. Quality control encompasses a series of measures and procedures aimed at maintaining the integrity of manufacturing processes and, by extension, the products themselves. Reliability, on the other hand, refers to the probability that a component or system will perform its required functions under stated conditions for a specified period of time. Together, these principles form the backbone of engineering practices that lead to durable, dependable products.

Quality control mechanisms are implemented at various stages of the manufacturing process, from the initial design phase through production to post-production inspection. These mechanisms include statistical process control (SPC), which utilizes statistical methods to monitor and control manufacturing processes. SPC can identify process variations that might lead to defects, allowing for corrective actions before the manufacturing process continues. Another critical aspect of quality control is the implementation of quality management systems (QMS), such as ISO 9001, which provide a framework for consistent quality in processes and products.

Reliability engineering focuses on minimizing system failures and enhancing product longevity. This discipline employs techniques such as failure mode and effects analysis (FMEA) to systematically evaluate potential failure points within a system and their possible impacts. Reliability can be quantified using metrics such as the mean time between failures (MTBF) for repairable systems and the mean time to failure (MTTF) for non-repairable systems. These metrics are crucial for understanding and improving the durability and performance of engineering designs.

The integration of **reliability-centered maintenance (RCM)** strategies is another critical aspect of ensuring component longevity. RCM prioritizes maintenance tasks based on the criticality of components and the consequences of their failure, thus optimizing maintenance efforts and resources. This approach not only enhances reliability but also contributes to overall safety and operational efficiency.

Quality and reliability testing are indispensable for validating the design and manufacturing processes. Testing methods range from accelerated life testing, which exposes products to elevated stress levels to predict their lifespan under normal conditions, to environmental testing, which assesses performance under varying environmental conditions. These tests provide empirical data that inform design improvements and ensure that products meet or exceed quality and reliability standards.

Incorporating quality and reliability principles into mechanical design and analysis requires a comprehensive understanding of both theoretical concepts and practical applications. Engineers must be adept at employing statistical analysis, understanding material properties, and anticipating the environmental and operational stresses that components will face. By prioritizing these principles, engineers can create products that not only meet the immediate needs of their users but also stand the test of time, thereby reflecting the true essence of engineering excellence.

Key Components in Mechanical Systems

Hydraulic, pneumatic, and electromechanical systems play pivotal roles in modern engineering, each comprising key components that enable their functionality and application across various industries. Understanding these components is essential for engineers aiming to design, analyze, and improve mechanical systems efficiently.

Hydraulic Systems rely on fluid power to perform work. Essential components include:

- **Pumps**: Convert mechanical energy into hydraulic energy by forcing fluid from the reservoir into the system.
- **Actuators** (Hydraulic cylinders and motors): Convert hydraulic energy back into mechanical energy to perform work.
- **Valves**: Control the flow and pressure within the system to direct the power to the desired actuators.
- **Reservoirs**: Store the hydraulic fluid necessary for the system's operation.
- **Filters**: Remove contaminants from the fluid to protect the system from wear and maintain fluid cleanliness.
- **Hoses and Tubes**: Transport hydraulic fluid between the system's components.

Pneumatic Systems, which utilize compressed air to transmit and control energy, include:

- **Compressors**: Generate the compressed air needed to power the system.
- **Actuators** (Pneumatic cylinders and motors): Perform the work by converting the energy of compressed air into mechanical force.

- **Valves**: Regulate the air flow and direction to the actuators.
- **Air Treatment Components**: Filters, regulators, and lubricators prepare the compressed air for use by removing impurities, controlling pressure, and adding lubrication to reduce wear.
- **Pipes and Fittings**: Distribute the compressed air to various parts of the system.

Electromechanical Systems combine electrical and mechanical processes and components, including:

- **Motors and Actuators**: Convert electrical energy into mechanical motion.
- **Sensors and Transducers**: Detect changes in physical conditions (like temperature, pressure, or position) and convert them into electrical signals.
- **Switches and Relays**: Control the operation of the system by opening or closing electrical circuits.
- **Power Supply Units (PSUs)**: Provide the electrical energy required to power the system.
- **Controllers (PLCs or microcontrollers)**: Programmable devices that direct the system's operation based on input from sensors and pre-defined algorithms.

Each of these systems has its unique advantages and applications. **Hydraulic systems** are favored for their high power density and precise control, making them ideal for heavy machinery and industrial applications. **Pneumatic systems** are preferred for their simplicity, cleanliness, and safety, especially in applications requiring a clean environment or explosive atmospheres. **Electromechanical systems** offer the best of both worlds, combining precise control with the flexibility of electrical systems, suitable for a wide range of applications from consumer electronics to robotics.

The selection of components within these systems is critical, not only for the system's initial performance but also for its longevity and reliability. Engineers must consider factors such as compatibility, environmental conditions, load requirements, and maintenance needs when designing and selecting components for hydraulic, pneumatic, and electromechanical systems. This careful consideration ensures that the system not only meets the required performance criteria but also operates efficiently and reliably over its intended lifespan, thereby maximizing productivity and minimizing downtime.

Engineering Drawing and GD&T Basics

Engineering drawings are the universal language of engineers, providing a comprehensive and detailed view of mechanical components and systems. The ability to accurately interpret these drawings is crucial for any engineer, as it enables the understanding of the design intent, dimensions, tolerances, and functionality of parts and assemblies. Geometric Dimensioning and

Tolerancing (GD&T) is a symbolic language used on engineering drawings to precisely define the geometry and allowable variation of a part. GD&T provides a clear and concise method for communicating complex design specifications and manufacturing constraints, ensuring that all team members have a common understanding of the part's requirements.

Dimensioning is the process of defining the size, location, orientation, and form of individual features on a part. Traditional dimensioning techniques include linear dimensions, which specify the distance between two points or the length of a feature, and angular dimensions, which define the angle between two lines or surfaces. Dimensions are typically accompanied by tolerances, which specify the allowable variation from the nominal dimension. Tolerances are critical for ensuring that parts fit together correctly and function as intended, without being overly restrictive or costly to manufacture.

GD&T goes beyond traditional dimensioning by providing a comprehensive system for defining the allowable geometric variation of part features. The GD&T system includes a set of standard symbols, rules, and definitions that describe the form, orientation, location, and runout of part features. Key GD&T concepts include:

- **Feature Control Frames**: A feature control frame is a rectangular box that contains the GD&T information for a feature. It specifies the geometric characteristic being controlled, the tolerance, the datum reference (if applicable), and any modifiers.

- **Datums**: Datums are theoretical planes, points, or axes on a part that serve as a reference for measuring and constraining geometric characteristics. Datums are essential for establishing the orientation and location of part features.

- **Geometric Characteristics**: GD&T defines fourteen geometric characteristics, such as flatness, straightness, circularity, cylindricity, perpendicularity, parallelism, and concentricity. Each characteristic has a specific symbol and represents a type of geometric control.

- **Tolerance Zones**: A tolerance zone is a three-dimensional region within which a feature's geometry must lie. GD&T allows for the specification of complex tolerance zones that can be more intuitive and flexible than traditional tolerances.

- **Modifiers**: Modifiers in GD&T provide additional information about the tolerance or condition of a feature. Common modifiers include maximum material condition (MMC), least material condition (LMC), and regardless of feature size (RFS).

Interpreting engineering drawings with GD&T requires a solid understanding of these concepts and the ability to apply them to real-world scenarios. Engineers must be able to read feature control frames, identify datum features and their hierarchy, and understand the implications of geometric tolerances on part manufacturability and assembly. Mastery of GD&T enables engineers to design more functional and cost-effective parts by optimizing tolerances, improving quality control, and facilitating clear communication among design, manufacturing, and quality assurance teams.

In practice, applying GD&T principles begins at the design stage, where engineers specify the geometric constraints of part features based on functional requirements and manufacturing capabilities. During manufacturing, machinists and quality inspectors use GD&T specifications to ensure parts meet the required standards. Finally, in assembly, engineers and technicians use GD&T to guide the fitting of components, ensuring that the final product operates as intended.

The interpretation of engineering drawings and GD&T is a critical skill for mechanical engineers, underpinning the successful design, manufacture, and assembly of mechanical systems. As such, engineers must continually refine their understanding and application of these standards to stay at the forefront of engineering excellence.

Made in United States
Orlando, FL
13 January 2025